Inspecting & Cleaning

Potable Water Storage

By Ron Perrin
© 2008

Print information available on the last page

Rev. date: 04/03/2019

To order additional copies of this book, contact:
Xlibris
1-888-795-4274
www.Xlibris.com
Orders@Xlibris.com

"A fundamental promise we must make to
our people is that the food they eat and the
water they drink are safe."

- President Bill Clinton, Safe Drinking Water Act
    - Reauthorization, August 6, 1996

CONTENTS

EPA GUIDELINES FOR FINISHED WATER

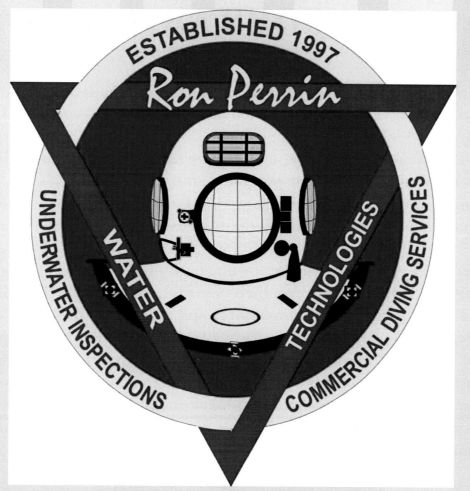

ALL Photos are Courtesy of Ron Perrin,
and *Ron Perrin Water Technologies* (RPWT)
Visit *www.ronperrin.com* for more details

Written by Ron Perrin

Edited by Wesley McCullough

Associate Editor Tony Dickerson

Illustrations by Robert Perrin

FORWARD

Since 1992 I have been involved in the inspection and cleaning of potable water systems. I have developed of innovative in-service cleaning and inspection equipment and techniques utilizing divers and remote controlled underwater equipment. These progressive underwater inspection methods have saved countless million's of gallons of potable water over the last decade. The water tank cleaning methods that have been developed have removed contaminants from hundreds of water systems, promoting and insuring public health.

This book started out as my own reference manual to allow me to have all the resources I needed to produce comprehensive inspection and cleaning reports. I have included additional notes and references to guide you through your own tank inspection or know what to ask for when hiring a contracted inspector. I review what is currently listed as contaminants in the nations' water supply, and highlight the importance of removing sediment that could be harboring and even allowing some of those contaminants to thrive and reproduce in your system. We will also review the most effective cleaning methods and what to look for when hiring a cleaning contractor.

Since September 11th 2001, the security of the nation's water supplies has risen to a much higher level. As directors, managers, employees or contractors we must all think about the security of our systems. We will review mandated requirements for systems that serve over 3,300 people in addition to practical steps that smaller systems may employ to secure their facilities.

Out of sight and out of mind, sediment in the bottom of your water storage tanks is never seen and rarely thought of. I hope this manual will help people understand what can be lurking in their water system, the importance of documenting proper inspections and the benefits of maintaining a clean and healthy system.

Ron Perrin
Ron Perrin Water Technologies
www.ronperrin.com

This book is dedicated to my father Charles B. Perrin.

He set a high standard for fatherhood, one that is not easily duplicated.
He allowed me to believe in myself.

Ron Perrin © 2008

Ron Perrin with his father Charles Perrin, sons Robert & Bradly Perrin

CHAPTER ONE

A brief history of water storage

and our understanding of the water in it

© Ron Perrin 2006

Since the beginning of time . . .

Since the beginning of time, man has needed water for his survival. With the first inventions of man, storing water to drink had to be on the top of his list. Egyptian

Paintings from as far back as the 15th century B.C. show us that they used a tank with what looks like wick siphons. Some scholars believe this was how they may have removed suspended solids.

Robert Perrin © 2008

Historically, water was considered clean if it was clear. The first know water filter was developed by Hippocrates, known as "Hippocrates Sleeve". It was a cloth bag to strain rainwater. This dates back to the 5th century B.C. Roman engineers created some of the first major public works projects with a water supply system that may have delivered 130 million gallons daily through aqueducts between 343 B.C. to 225 A.D.

Little was done to improve water systems or treatment for over the next 1,000 years. The theory of clear equals clean stood until the 1800's.

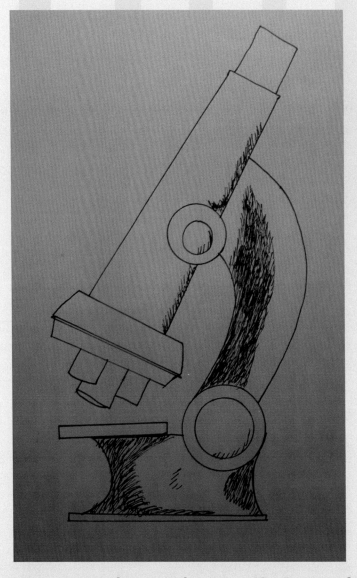

The microscope was invented in 1680 by Anton Van Leeuwenhoek and discovered microscopic organisms. By 1685 Italian physician Lu Antonio Porzio designed the first multiple filter. These two unrelated events were to play important parts in the future of water treatment. Porzio's filter used sedimentation and straining followed by sand filtration. This system had two compartments one with downward flow, and one with upward flow. In 1804 Paisley, Scotland was the first city to deliver water to an entire town. Within three years, filtered water was even piped directly to customers in Glasgow, Scotland.

In **1806** a large scale sand and charcoal water treatment plant began operating in Paris.

America has relied on several different processes of water treatment since the **1870's,** when Dr. Robert Koch and Dr. Joseph Lister proved micro-organisms in water supplies can cause disease.

After the Civil War, the United States became a leader in water works by further developing filtration.

In **1881** the American Water Works Association was formed at a meeting of 22 water works professionals at Washington University.

Significant improvements to water treatment in the latter part of the **1800's** included the development of rapid sand and slow sand filters, and the first applications of chlorine and ozone for disinfection. At the turn of the century, chlorination became the most popular method in the United States and numbers of typhoid dysentery and cholera cases were greatly reduced.

Jersey City Water Works became the first utility in America to use sodium hypochlorite for disinfection in **1908**, and the Bubbly Creek plant in Chicago instituted regular chlorine disinfection.

In **1918** the Texas Water Utilities Association was formed at Texas A&M University making it the oldest State Association in the United States.

By the **1920's,** the use of filtration and chlorination had virtually eliminated epidemics of major waterborne diseases from the American landscape. These two decades also saw the development of dissolved air flotation, early membrane filters, flock blanket sedimentation, and the solids-contact clarifier.

A major step in the development of desalination technology arrived during World War II when various military establishments in arid areas required water to supply their troops.

The U.S. Public Health Service was adopted in **1942**. In **1957** the first set of drinking water standards, and the membrane filter process for bacteriological analysis was approved.

Water systems have gone from 19,000 in **1960** to over 66,000 municipal systems, producing over 38 billion gallons of water per day for domestic and public use. In addition to this over 13 million households get their water from their own private wells and are responsible for treating and pumping the water themselves in the United States. We use an average of 27 gallons per person per day with up to 50% of this consumption going to watering lawns.

In **1969** the Texas Rural Water Association was founded. The National Rural Water Association now claims to have the largest utility membership, currently serving over 24,550 Water & Wastewater Utilities.

In **1974** the Safe Drinking Water Act (THE SDWA) was originally passed by Congress. The ACT was put in place to protect public health by regulating the nation's public drinking water supply. The law was amended in 1986 and 1996 and requires many actions to protect drinking water and its sources: rivers, lakes, reservoirs, springs, and ground water wells.

In **1990**, the EPA Science Advisory Board (SAB) cited drinking water contamination as one of the most important environmental risks and indicated that disease causing microbial contaminants (i.e., bacteria, protozoa, and viruses) is probably the greatest remaining health risk management challenge for drinking water suppliers (USEPA/SAB, 1990).

In **1996**, amendments greatly enhanced the existing law by recognizing source water protection, operator training, funding for water system improvements, and public information as important components of safe drinking water.

In **1999**, the EPA, along with the Center for Disease Control and Prevention, co-released a paper titled "Guidance for People with Severely Weakened Immune Systems". *Cryptosporidium* is a parasite commonly found in lakes and rivers, especially when the water is contaminated with sewage and animal wastes. *Cryptosporidium* is very resistant to disinfection, and <u>even a well-operated water treatment system</u> cannot ensure that drinking water will be completely free of this parasite.

In **2000** the Office of Ground Water and Drinking Water issued a paper on ARSENIC OCCURRENCE IN PUBLIC DRINKING WATER SUPPLIES

In **2002** the Office of Water (4601M), Office of Ground Water and Drinking Water issued a Distribution System Issue Paper titled

"Finished Water Storage Facilities"

This paper outlined contaminants found in potable water storage and the importance of inspection & cleaning. Pathogen contamination and microbial growth in water storage systems is covered extensively in this paper. Tank inspections using robotic devices or divers and regular tank cleaning are also covered in this paper. You will find this document reprinted in its entirety in the back of the book.

In **2005** A national assessment of tap water quality conducted by the *Environmental Working Group* found that tap water in 42 states is contaminated with more than 140 unregulated chemicals that lack safety standards. The assessment took

Two and a half years. Analysis of more than 22 million tap water quality test detected 260 contaminates in water served to the public.

In **2006** Office of Water (4601M), Office of Ground Water and Drinking Water Total Coli form Rule Issue Paper entitled "Inorganic Contaminant Accumulation in Potable Water Distribution Systems"

This paper lists the regulated inorganic compounds including:
Antimony, arsenic, barium, beryllium, cadmium, chromium, copper,
Cyanide, Fluoride, Lead, Mercury, Nitrite and nitrate, Selenium, and Thallium.

Unregulated inorganic compounds are also listed including:
Aluminum, Manganese, Nickel, Silver, Vanadium, Zinc and rare-earth elements (REE) are scandium (Sc), yttrium (Y), and the 15 elements on the periodic table between and including lanthanum (La) and lutetium (Lu) including dysprosium (Dy). These elements are all trivalent and have very similar chemical properties

CHAPTER 2

RPWT © 2008 Leaning tower near Amarillo, Texas.

The EPA and Potable water

The mission of the Environmental Protection Agency is to protect human health and the environment. Since 1970, EPA has been working for a cleaner, healthier environment for the American people.

The EPA employs 18,000 people across the country, including our headquarters offices in Washington, DC, 10 regional offices, and more than a dozen labs. Their staff is highly educated and technically trained; more than half are engineers, scientists, and policy analysts. In addition, a large number of employees are legal, public affairs, financial, information management and computer specialists. The EPA is led by the Administrator, who is appointed by the President of the United States.

Most Americans get their drinking water from large scale municipal water systems that rely on surface water sources such as rivers, lakes and reservoirs. However, millions of Americans depend on private water sources such as wells and aquifers. In either case, the United States enjoys one of the cleanest drinking water supplies in the world. The EPA regulates the quality of the nation's drinking water by issuing and enforcing safe drinking water standards. EPA also protects the nation's drinking water by safeguarding our watersheds and regulating the release of pollutants into the environment. In partnership with local authorities and community groups, the Agency encourages water conservation. The EPA also works with these partners to develop contingency plans for source contamination and other water emergencies.

Where does your drinking water come from? How do you know if your drinking water is safe? How can you protect it? What can you do if there's a problem with your drinking water?

To help answer these — and other — questions, the U.S. Environmental Protection Agency prepared "Water on Tap: What You Need to Know" in both English and a Chinese translation.

If you do not have a copy of "Water on Tap" get one now.
To order a copy of the English version by mail:
Water On Tap, #634D (Please allow 4 to 6 weeks for delivery)
Consumer Information Center
Pueblo, CO 81009

The **S**afe **D**rinking **W**ater **A**ct (SDWA) was originally passed by Congress in 1974 to protect public health by regulating the nation's public drinking water supply. The law was amended in 1986 and again in 1996 and requires many actions to protect drinking water and its sources: rivers, lakes, reservoirs, springs, and ground water wells. (THE SDWA does not regulate private wells which serve fewer than 25 individuals.)

The SDWA authorizes the **U**nited **S**tates **E**nvironmental **P**rotection **A**gency (EPA) to set national health-based standards for drinking water to protect against both naturally-occurring and man-made contaminants that may be found in drinking water. The EPA, states, and water systems then work together to make sure that these standards are met.

Millions of Americans receive high quality drinking water every day from their public water systems, (which may be publicly or privately owned). Nonetheless, drinking water safety cannot be taken for granted. There are a number of threats to drinking water: improperly disposed of chemicals, animal wastes, pesticides, human wastes, wastes injected deep underground, and naturally-occurring substances can all contaminate drinking water. Likewise, drinking water that is not properly treated or disinfected, or which travels through an improperly maintained distribution system, may also pose a health risk.

Originally, the SDWA focused primarily on treatment as the means of providing safe drinking water at the tap. The 1996 amendments greatly enhanced the existing law by recognizing source water protection, operator training, funding for water system improvements, and public information as important components of safe drinking water. This approach ensures the quality of drinking water by protecting it from source to tap.

Roles and responsibilities

The SDWA applies to every public water system in the United States. There are currently more than 160,000 public water systems providing water to almost all Americans at some time in their lives. The responsibility for making sure these public water systems provide safe drinking water is divided among the EPA, states, tribes, water systems, and the public. The SDWA provides a framework in which these parties work together to protect this valuable resource.

The EPA sets national standards for drinking water based on sound science to protect against health risks, considering available technology and costs. These National Primary Drinking Water Regulations set enforceable maximum contaminant levels for particular contaminants in drinking water and required ways to treat water to remove contaminants. Each standard also includes requirements for water systems to test for contaminants in the water to make sure standards are achieved. In addition to setting these standards, The U.S EPA provides guidance, assistance, and public information about drinking water, collects drinking water data, and oversees state drinking water programs.

The most direct oversight of water systems is conducted by state drinking water programs. States can apply to The EPA for "primacy," the authority to implement the SDWA within their jurisdictions, if they can show that they will adopt standards at least as stringent as The EPA's and make sure water systems meet these standards. All states and territories, except Wyoming and the District of Columbia, have received primacy. While no Indian tribe has yet applied for and received primacy, four tribes currently receive "treatment as a state" status, and are eligible for primacy. States, or the EPA acting as a primacy agent, make sure water systems test for contaminants, review plans for water system improvements, conduct on-site inspections and sanitary surveys, provide training and technical assistance, and take action against water systems not meeting standards.

To ensure that drinking water is safe, the SDWA sets up multiple barriers against pollution. These barriers include: source water protection, treatment, distribution system integrity, and public information. Public water systems are responsible for ensuring that contaminants in tap water do not exceed the standards. Water systems treat the water, and must test their water frequently for specified contaminants and report the results to states. If a water system is not meeting these standards, it is the water supplier's responsibility to notify its customers. Many water suppliers now are also required to prepare annual reports for their customers. The public is responsible for helping local water suppliers to set priorities, make decisions on funding and system improvements, and establish programs to protect drinking water sources. Water systems across the nation rely on citizen advisory committees, rate boards, volunteers, and civic leaders to actively protect this resource in every community in America.

Protection and prevention

Essential components of safe drinking water include protection and prevention. States and water suppliers must conduct assessments of water sources to see where they may be vulnerable to contamination. Water systems may also voluntarily adopt programs to protect their watershed or wellhead and states can use legal authorities from other laws to prevent pollution. The SDWA mandates that states have programs to certify water system operators and make sure that new water systems have the technical, financial, and managerial capacity to provide safe drinking water. THE SDWA also sets a framework for the Underground Injection Control (UIC) program to control the injection of wastes into ground water. The EPA and states implement the UIC program, which sets standards for safe waste injection practices and bans certain types of injection altogether. All of these programs help prevent the contamination of drinking water.

Setting national drinking water standards

The EPA sets national standards for tap water which help ensure consistent quality in our nation's water supply. The EPA prioritizes contaminants for potential regulation based on risk and how often they occur in water supplies. (To aid in this effort, certain water systems monitor for the presence of contaminants for which no national standards currently exist and collect information on their occurrence). The EPA sets a health goal based on risk (including risks to the most sensitive people, e.g., infants, children, pregnant women, the elderly, and the immune-compromised). The EPA then sets a legal limit for the contaminant in drinking water or a required treatment technique. This limit or treatment technique is set to be as close to the health goal as feasible. The EPA also performs a cost-benefit analysis and obtains input from interested parties when setting standards. The EPA is currently evaluating the risks from several specific health concerns, including: microbial contaminants (e.g., Cryptosporidium), the byproducts of drinking water disinfection, radon, arsenic, and water systems that don't currently disinfect their water but get it from a potentially vulnerable ground water source

Funding and assistance

The EPA provides grants to implement state drinking water programs, and to help each state set up a special fund to assist public water systems in financing the costs of improvements (called the drinking water state revolving fund). Small water systems are given special consideration, since small systems may have a more difficult time paying for system improvements due to their smaller customer base. Accordingly, The EPA and states provide them with extra assistance (including training and funding) as well as allowing, on a case-by-case basis, alternate water treatments that are less expensive, but still protective of public health

Compliance and enforcement

National drinking water standards are legally enforceable, which means that both The EPA and states can take enforcement actions against water systems not meeting safety standards. The EPA and states may issue administrative orders, take legal actions, or fine utilities. The EPA and states also work to increase water systems, understanding of, and compliance with, these standards.

Public information

The SDWA recognizes that since everyone drinks water, everyone has the right to know what's in it and where it comes from. All water suppliers must notify consumers quickly when there is a serious problem with water quality. Water systems serving the same people year-round must provide annual consumer confidence reports on the source and quality of their tap water. States and The EPA must prepare annual summary reports of water system compliance with drinking water safety standards and make these reports available to the public. The public must have a chance to be involved in developing source water assessment programs, state plans to use drinking water state revolving loan funds, state capacity development plans, and state operator certification programs

1996 The SDWA Amendment Highlights

Consumer Confidence Reports:

All community water systems must prepare and distribute annual reports about the water they provide, including information on detected contaminants, possible health effects, and the water's source.

Cost-Benefit Analysis:

The EPA must conduct a thorough cost-benefit analysis for every new standard to determine whether the benefits of a drinking water standard justify the costs.

Drinking Water State Revolving Fund:

States can use this fund to help water systems make infrastructure or management improvements or to help systems assess and protect their source water.

Microbial Contaminants and Disinfection Byproducts:

The EPA is required to strengthen protection for microbial contaminants, including Cryptosporidium, while strengthening control over the byproducts of chemical disinfection. The Stage 1 Disinfectants and Disinfection Byproducts Rule and the Interim Enhanced Surface Water Treatment Rule together address these risks.

Operator Certification:

Water system operators must be certified to ensure that systems are operated safely. The EPA issued guidelines in February 1999 specifying minimum standards for the certification and recertification of the operators of community and non-transient, non-community water systems. These guidelines apply to state Operator Certification Programs. All States are currently implementing EPA-approved operator certification programs.

Public Information & Consultation:

The SDWA emphasizes that consumers have a right to know what is in their drinking water, where it comes from, how it is treated, and how to help protect it. The EPA distributes public information materials (through its Safe Drinking Water Hotline, Safewater web site, and Water Resource Center) and holds public meetings, working with states, tribes, water systems, and environmental and civic groups, to encourage public involvement.

Small Water Systems:

Small water systems are given special consideration and resources under The SDWA, to make sure they have the managerial, financial, and technical ability to comply with drinking water standards.

Source Water Assessment Programs:

Every state must conduct an assessment of its sources of drinking water (rivers, lakes, reservoirs, springs, and ground water wells) to identify significant potential sources of contamination and to determine how susceptible the sources are to these threats.

For more information on the EPA rules, programs and grants go to: http://www.epa.gov/safewater/The SDWA/basicinformation.html

CHAPTER THREE

STATE GUIDELINES ON POTABLE WATER STORAGE

STATE RULES

States that Require Potable Water Tanks to be Inspected on regular intervals.

According to the EPA, States that do have recommendations are Alabama (5 years), Arkansas (2 years), Missouri (5 years), New Hampshire (5 years), Ohio (5 years), Rhode Island (external once per year; internal, every five years), Texas (annually), and Wisconsin (5 years). The most current EPA records I could find on this subject were published in 2002. I personally contacted individual states and here are the responses I received:

Alabama

Water Supply Branch
Dept. of Environmental Management
P.O. Box 301463
1400 Coliseum Blvd.
Montgomery, AL 36130-1463
Phone: 334-271-7773

ADEM Admin. Code 335-7-7-.04 requires:

"All public water systems are to have their water storage tanks inspected on regular intervals. Any deficiencies identified during the inspection should be repaired in a timely manner."

The regulations leave it up to the water system to establish the frequency of tank inspection and the type of inspection performed. I hope this answers your question. Alabama regulations can be downloaded under Division 7-Water Supply Program at <u>www.adem.state.al.us</u> ."

We found it Here: <u>http://www.adem.state.al.us/Regulations/regulations.htm</u> (effective date—December 12, 2005) Page 106

Connecticut

Drinking Water Division
Dept. of Public Health
MS-51WAT
P.O. Box 340308
Hartford, CT 06134-0308
860-509-7333

The Law Unit at the Connecticut State Library sent me this response; "Our laws and regulations are not so straightforward. I have attached links to sections of the state statutes and regulations concerning public water supply plans. These plans must be approved by the Department of Public Health. To be approved, a plan must have a description of the company's testing, but I did not see any greater detail of what that would involve.

I am also attaching contact information for the Department of Public Health.
Staff there should be able to answer your question. http://www.dph.state.ct.us/BRS/water/Utility/utility.htm http://www.dph.state.ct.us/BRS/water/Contact_Info/contact_info.htm http://www.dph.state.ct.us/phc/docs/151_Water_Supply_Plans.doc http://www.dph.state.ct.us/phc/docs/31_Public_Drinking_Water_Qu.doc http://www.dph.state.ct.us/phc/docs/152_Civil_Penalties_for_Vio.doc

Minnesota

Drinking Water Protection Section
Dept. of Health
121 East Seventh Place
P.O. Box 64975
St. Paul, MN 55164-0975
612-215-0770

The Minnesota Department of Health recommends to our regulated community that they regularly inspect and provide maintenance for potable water storage tanks. The regular inspections should make use of both remote inspection equipment, and physical on-site inspection by a qualified inspector in the tank. The inspections should be conducted in accordance with the American Water Works Association (AWWA) standards, and other industry standards. The frequency of the inspections and cleaning is left up to the owner of the system, unless there are frequent water quality issues that are reported or observed.

Any equipment that is used in the inspection and maintenance of potable water storage tanks must not negatively affect the water quality of the stored potable water. All equipment that is used must be dedicated to use only in potable water installations, and shall be thoroughly disinfected with a high concentration chlorine solution prior to being placed into service. Any material that is removed from the tank during cleaning and maintenance must be disposed of in an approved manner.

Evaluations and interpretations of the information gathered at the time of inspection should be done by a **"qualified, certified engineer that has been properly trained in metal corrosion and coatings methods. The engineer should also be registered in the State of Minnesota."**

Rhode Island

Office of Drinking Water Quality
Dept. of Health
3 Capitol Hill
Room 209
Providence, RI 02911
401-222-6867

Water tanks must be inspected, but there is no regulation that specifically states this. Rather, if tanks were not being inspected, then it would be a matter for follow-up when discovered during the sanitary survey.

States that Require Potable Water Tanks to be Inspected annually.

Pennsylvania

Facility Permits Section
Division of Planning and Permits
P.O. Box 8774
Harrisburg, PA 17105-8774

Pennsylvania's Safe Drinking Water regulations are found in Chapter 109 of Title 25 of the Pennsylvania Code. The regulation can be found at:

http://www.pacode.com/secure/data/025/025toc.html

Paragraph (a) of Section 109.705 of the regulation requires community water suppliers to conduct a sanitary survey of the water system at least annually. Subparagraph (4) requires an evaluation of finished water storage facilities.

Kevin S. McLeary, P.E is with the Division of Planning and Permits. McLeary wrote the following: Although the regulation requires water systems to conduct sanitary surveys, it provides no methodology for conducting them. EPA has published a Guidance Manual for Conducting Sanitary Surveys of Public Water Systems. The EPA manual recommends that an inspector examine the design criteria of the storage tanks to meet the demands of the system and retain structural integrity, and that the entire tank and all appurtenances be inspected by qualified personnel with the results documented in a written report. Pennsylvania is not permitted to "regulate by guidance." So, while the recommendations in the manual are highly encouraged, they cannot be considered to be absolute requirements.

Texas

Water Supply Division (MC-155)
Texas Commission on Environmental Quality
P.O. Box 13087
Austin, TX 78711-3087
512-239-4691

http://www.tceq.state.tx.us/rules/index.html

Texas Administrative Code

Title 30 ENVIRONMENTAL QUALITY

Part 1 TEXAS COMMISSION ON ENVIRONMENTAL QUALITY

CHAPTER 290 PUBLIC DRINKING WATER

SUBCHAPTER D RULES AND REGULATIONS FOR PUBLIC WATER SYSTEMS

RULE §290.46 Minimum Acceptable Operating Practices

for Public Drinking Water Systems

(m) Maintenance and housekeeping. The maintenance and housekeeping practices used by a public water system shall ensure the good working condition and general appearance of the system's facilities and equipment. The grounds and facilities shall be maintained in a manner so as to minimize the possibility of the harboring of rodents, insects, and other disease vectors, and in such a way as to prevent other conditions that might cause the contamination of the water.

(1) Each of the system's ground, elevated, and pressure tanks shall be inspected annually by water system personnel or a contracted inspection service.

(A) Ground and elevated storage tank inspections must determine that the vents are in place and properly screened, the roof hatches closed and locked, flap valves and gaskets provide adequate protection against insects, rodents, and other vermin, the interior and exterior coating systems are continuing to provide adequate protection to all metal surfaces, and the tank remains in a watertight condition.

(B) Pressure tank inspections must determine that the pressure release device and pressure gauge are working properly, the air-water ratio is being maintained at the proper level, the exterior coating systems are continuing to provide adequate protection to all metal surfaces, and the tank remains in watertight condition. Pressure tanks provided with an inspection port must have the interior surface inspected **every five years**.

(C) All tanks shall be inspected annually to determine that instrumentation and controls are working properly.

With all that said under Section 290.46 you still do not have the information you need to perform an annual inspection in Texas until you back up and read rule 290.43:

RULE §290.43 Water Storage

(a) Capacity. The minimum clearwell, storage tank, and pressure maintenance capacity shall be governed by the requirements in §290.45 of this title (relating to Minimum Water System Capacity Requirements).

(b) Location of clearwells, standpipes, and ground storage and elevated tanks.

(1) No public water supply elevated storage or ground storage tank shall be located within 500 feet of any municipal or industrial sewage treatment plant or any land which is spray irrigated with treated sewage effluent or sludge disposal.

(2) Insofar as possible, clearwells or treated water tanks shall not be located under any part of any buildings and, when possible, shall be constructed partially or wholly above ground.

(3) No storage tank or clearwell located below ground level is allowed within 50 feet of a sanitary sewer or septic tank. However, if the sanitary sewers are constructed of 150 pounds per square inch (psi) pressure-rated pipe with pressure-tested, watertight joints as used in water main construction, the minimum separation distance is ten feet.

(4) No storage tank or clearwell located below ground level is allowed within 150 feet of a septic tank soil absorption system.

(c) Design and construction of clearwells, standpipes, ground storage tanks, and elevated tanks. All facilities for potable water storage shall be covered and designed, fabricated, erected, tested, and disinfected in strict accordance with current American Water Works Association (AWWA) standards and shall be provided with the minimum number, size and type of roof vents, man ways, drains, sample connections, access ladders, overflows, liquid level indicators, and other appurtenances as specified in these rules. Bolted tanks shall be designed, fabricated, erected, and tested in strict accordance with current AWWA Standard D103. The roof of all tanks shall be designed and erected so that no water ponds at any point on the roof and, in addition, no area of the roof shall have a slope of less than 0.75 inch per foot.

(1) Roof vents shall be gooseneck or roof ventilator and be designed by the engineer based on the maximum outflow from the tank. Vents shall be installed in strict accordance with current AWWA standards and shall be equipped with approved screens to prevent entry of animals, birds, insects and heavy air contaminants. Screens shall be fabricated of corrosion-resistant material and shall be 16-mesh or finer. Screens shall be securely

clamped in place with stainless or galvanized bands or wires and shall be designed to withstand winds of not less than tank design criteria (unless specified otherwise by the engineer).

(2) All roof openings shall be designed in accordance with current AWWA standards. If an alternate 30 inch diameter access opening is not provided in a storage tank, the primary roof access opening shall not be less than 30 inches in diameter. Other roof openings required only for ventilating purposes during cleaning, repairing or painting operations shall be not less than 24 inches in diameter or as specified by the licensed professional engineer. An existing tank without a 30-inch in diameter access opening must be modified to meet this requirement when major repair or maintenance is performed on the tank. Each access opening shall have a raised curbing at least four inches in height with a lockable cover that overlaps the curbing at least two inches in a downward direction. Where necessary, a gasket shall be used to make a positive seal when the hatch is closed. All hatches shall remain locked except during inspections and maintenance.

(3) Overflows shall be designed in strict accordance with current AWWA standards and shall terminate with a gravity-hinged and weighted cover. The cover shall fit tightly with no gap over 1/16 inch. If the overflow terminates at any point other than the ground level, it shall be located near enough and at a position accessible from a ladder or the balcony for inspection purposes. The overflow(s) shall be sized to handle the maximum possible fill rate without exceeding the capacity of the overflow(s). The discharge opening of the overflow(s) shall be above the surface of the ground and shall not be subject to submergence.

(4) All clearwells and water storage tanks shall have a liquid level indicator located at the tank site. The indicator can be a float with a moving target, an ultrasonic level indicator, or a pressure gauge calibrated in feet of water. If an elevated tank or standpipe has a float with moving target indicator, it must also have a pressure indicator located at ground level. Pressure gauges must not be less than three inches in diameter and calibrated at not more than two-foot intervals. Remote reading gauges at the owner's treatment plant or pumping station will not eliminate the requirement for a gauge at the tank site unless the tank is located at the plant or station.

(5) Inlet and outlet connections shall be located so as to prevent short-circuiting or stagnation of water. Clearwells used for disinfectant contact time shall be appropriately baffled.

(6) Clearwells and potable water storage tanks shall be thoroughly tight against leakage, shall be located above the groundwater table, and shall have no walls in common

with any other plant units containing water in the process of treatment. All associated appurtenances including valves, pipes, and fittings shall be tight against leakage.

(7) Each clearwell or potable water storage tank shall be provided with a means of removing accumulated silt and deposits at all low points in the bottom of the tank. Drains shall not be connected to any waste or sewage disposal system and shall be constructed so that they are not a potential agent in the contamination of the stored water.

(8) All clearwells, ground storage tanks, standpipes, and elevated tanks shall be painted, disinfected, and maintained in strict accordance with current AWWA standards. However, no temporary coatings, wax grease coatings, or coating materials containing lead will be allowed. No other coatings will be allowed which are not approved for use (as a contact surface with potable water) by the EPA, National Sanitation Foundation (NSF), or United States Food and Drug Administration (FDA). All newly installed coatings must conform to American National Standards Institute/National Sanitation Foundation (ANSI/ NSF) Standard 61 and must be certified by an organization accredited by ANSI.

(9) No tanks or containers shall be used to store potable water that has previously been used for any non-potable purpose. Where a used tank is proposed for use, a letter from the previous owner or owners must be submitted to the Commission which states the use of the tank.

(10) Access manways in the riser pipe, shell area, access tube, bowl area or any other location opening directly into the water compartment shall be located in strict accordance with current AWWA standards. These openings shall not be less than 24 inches in diameter. However, in the case of a riser pipe or access tube of 36 inches in diameter or smaller, the access manway may be 18 inches times 24 inches with the vertical dimension not less than 24 inches. The primary access manway in the lower ring or section of a ground storage tank shall be not less than 30 inches in diameter. Where necessary, for any access manway which allows direct access to the water compartment, a gasket shall be used to make a positive seal when the access manway is closed.

(d) Design and construction of pressure (hydropneumatic) tanks. All hydropneumatic tanks must be located wholly above grade and must be of steel construction with welded seams except as provided in paragraph (8) of this subsection.

(1) Metal thickness for pressure tanks shall be sufficient to withstand the highest expected working pressures with a four to one factor of safety. Tanks of 1,000 gallons capacity or larger must meet the standards of the American Society of Mechanical Engineers (ASME) Section VIII, Division 1 Codes and Construction Regulations and must have an access port for periodic inspections. An ASME name plate must be permanently

attached to those tanks. Tanks installed before July 1, 1988, are exempt from the ASME coding requirement, but all new installations must meet this regulation. Exempt tanks can be relocated within a system but cannot be relocated to another system.

(2) All pressure tanks shall be provided with a pressure release device and an easily readable pressure gauge.

(3) Facilities shall be provided for maintaining the air-water-volume at the design water level and working pressure. Air injection lines must be equipped with filters or other devices to prevent compressor lubricants and other contaminants from entering the pressure tank. A device to readily determine air-water-volume must be provided for all tanks greater than 1,000 gallon capacity. Galvanized tanks which are not provided with the necessary fittings and which were installed before July 1, 1988 shall be exempt from this requirement.

(4) Protective paint or coating shall be applied to the inside portion of any pressure tank. The coating shall be as specified in subsection (c)(8) of this section.

(5) No pressure tank that has been used to store any material other than potable water may be used in a public water system. A letter from the previous owner or owners must be provided as specified in subsection (c)(9) of this section.

(6) Pressure tank installations should be equipped with slow closing valves and time delay pump controls to eliminate water hammer and reduce the chance of tank failure.

(7) All associated appurtenances including valves, pipes and fittings connected to pressure tanks shall be thoroughly tight against leakage.

(8) Where seamless fiberglass tanks are utilized, they shall not exceed 300 gallons in capacity.

(9) No more than three pressure tanks shall be installed at any one site without the prior approval of the executive director.

(e) Facility security. All potable water storage tanks and pressure maintenance facilities must be installed in a lockable building that is designed to prevent intruder access or enclosed by an intruder-resistant fence with lockable gates. Pedestal-type elevated storage tanks with lockable doors and without external ladders are exempt from this requirement. The gates and doors must be kept locked whenever the facility is unattended.

(f) Service pumps. Service pump installations taking suction from storage tanks shall provide automatic low water level cutoff devices to prevent damage to the pumps. The service pump circuitry shall also resume pumping automatically once the minimum water level is reached in the tank.

States that Require Potable storage tanks to be inspected once every five years:

Florida

Florida Department of Environmental Protection
Drinking Water Program (MS 3520)
2600 Blair Stone Road
Tallahassee, Florida 32399-2400 john.r.sowerby@dep.state.fl.us
Telephone: (850) 245-8637
Fax: (850) 245-8669

Florida Administrative Code (F.A.C.)

Subsection 62-555.350 Operation and Maintenance of Public Water Systems.

(1) Suppliers of water shall operate and maintain their public water systems so as to comply with applicable standards in Chapter 62-550, F.A.C., and requirements in this chapter.

(2) Suppliers of water shall keep all necessary public water system components in operation and shall maintain such components in good operating condition so the components function as intended. Preventive maintenance on electrical or mechanical equipment – including exercising of auxiliary power sources, checking the calibration of finished-drinking-water meters at treatment plants, testing of air or pressure relief valves for hydropneumatic tanks, and exercising of isolation valves – shall be performed in accordance with the equipment manufacturer's recommendations or in accordance with a written preventive maintenance program established by the supplier of water; however, in no case shall auxiliary power sources be run under load less frequently than monthly. **Accumulated sludge and biogrowths shall be cleaned routinely (i.e., at least annually) from all treatment facilities that are in contact with raw, partially treated, or finished drinking water** and that are not specifically designed to collect sludge or support a biogrowth; and blistering, chipped, or cracked coatings and linings on treatment or storage facilities in contact with raw, partially treated, or finished

drinking water shall be rehabilitated or repaired. **Finished-drinking-water storage tanks, including conventional hydropneumatic tanks with an access manhole but excluding bladder—or diaphragm-type hydropneumatic tanks without an access manhole, shall be checked at least annually to ensure that hatches are closed and screens are in place; shall be cleaned at least once every five years to remove biogrowths, calcium or iron/manganese deposits, and sludge from inside the tanks; and shall be inspected for structural and coating integrity at least once every five years by personnel under the responsible charge of a professional engineer licensed in Florida.**

Additionally, paragraph 62-555.350(12)(c), F.A.C., states the following:

"All suppliers of water shall keep records documenting that their finished-water-storage tanks, including hydropneumatic tanks with an access manhole but excluding bladder—or diaphragm-type hydropneumatic tanks without an access manhole, have been cleaned and inspected during the past five years in accordance with subsection 62-555.350(2), F.A.C . . . ."

You can view or download a copy of Chapter 62-555, F.A.C., at the following Florida Department of Environmental Protection web page: http://www.dep.state.fl.us/legal/Rules/rulelistnum.htm.

Reference: John R. Sowerby, P.E.

Florida also gets my VISIONARY AWARD for being the first state to require potable water storage cleaning. The way I understand this rule is that all water storage facilities in Florida must me cleaned at least every five years and that tanks like clearwells at water plants must be cleaned at least every year:

> Accumulated sludge and biogrowths shall be cleaned routinely (i.e., at least annually) from all treatment facilities that are in contact with raw, partially treated, or finished drinking water –

Florida has definitely set the standard that many states are sure to follow. Due to constant rule changes and updates this chapter is the one that will be out of date long before the rest. Check with your state to see what the current rules are and if they know of any upcoming changes.

Nebraska

Nebraska Dept. of HHS Regulation & Licensure
301 Centennial Mall South
P.O. Box 95007, 3rd Floor
Lincoln, NE 68509-5007
402-471-2541

In Nebraska, Title 179 Regulations Governing Public Water Supply Systems Chapter 22-008 item 1 requires that all Community and Non-Transient Non-Community public water systems which have storage facilities that are equipped for accessibility must have their storage facility inspected, and if necessary cleaned, no less often than every five (5) years.

Although not specified in regulation all underwater/diving type inspections and cleaning of storage facilities would have to be done under sterile/disinfected conditions. Sampling before the inspection/cleaning activity is highly recommended and sampling following the maintenance is required to be sure there was no contamination introduced into the facility during the inspection/cleaning process. The sampling of water from the storage facility must consist of a minimum of 2 bacteriological samples collected at least 24 hours apart. In order for the facility sampling to be "acceptable" you would have to obtain 2 consecutive "clean, coli form absent" samples. The Department would recommend that the storage facility be taken out of service during maintenance. If the storage facility must be used to supply water to the distribution system during maintenance, a disinfectant should be added to the source water so that all water entering the distribution system through the storage facility has received the benefit of a disinfectant.

Additionally, Title 179 Chapter 2-007.02A6 requires that plans and specifications be submitted for repairs to existing storage facilities including interior linings, painting or coatings. Other types of physical repair would also be covered under this requirement.

These regulations may be found at the following website:
http://www.hhs.state.ne.us/reg/t179.htm

New Hampshire

Water Supply Engineering Bureau
Dept. of Environmental Services
P.O. Box 95
6 Hazen Drive
Concord, NH 03302-0095
603-271-3139

Env-Ws 361.08 Storage Tank Inspection and Maintenance.

At Least once every 5 years, the water system owner shall cause the insides of each water storage tank used by the system to be inspected and evaluated for structural strength, corrosion and other factors related to water quality.

Metal surfaces shall be painted or otherwise protected as determined by the system owner, using materials in accordance with Env-Ws 305.

Visit   http://www.des.nh.gov/rulemaking/adopted2005/Env-Ws_360-362.pdf   for more NH rules.

South Carolina

Bureau of Water
Dept. of Health & Environmental Control
2600 Bull Street
Columbia, SC 29201
803-734-5300

When I contacted South Carolina, Richard Welch, Manager with the Drinking Water & Recreational Waters Compliance Section at the SC Dept of Health & Environmental Control stated that they do NOT have any regulations on when a tank should be cleaned. He is the manager of the compliance section and they are responsible for the sanitary survey program for all drinking water facilities across the state. Water storage tanks are one of the things that his inspectors evaluate during these inspections. While they don't have regulations, they can and do make recommendations as to when they should be evaluated and cleaned.

He advises water system's to climb all their tanks at least once per year. He also tells them that they should do a more in depth check of them every three to five years.

Tennessee

Division of Water Supply
Dept. of Environment & Conservation
401 Church Street
L & C Tower, 6th Floor
Nashville, TN 37243-1549
615-532-0191

Tanks in Tennessee have to be inspected every 5 years in accordance with regulation 1200-5-1-.17(33) which states:

(33) All public water systems shall properly maintain their distribution system finished water storage tanks. Each community water system shall establish and maintain a maintenance file on each of its finished water and distribution storage tanks. These maintenance files must be available for inspection by the Department's personnel. These files must include the dates and results of all routine water storage tank inspections by system personnel, any reports of detailed professional inspections of the water storage tanks by contractor personnel, dates and details of routine tank cleanings and surface flushings, and dates and details of all tank maintenance activities.

The tank inspection records shall include dates of the inspections; the sanitary, coating and structural conditions of the tank; and all recommendations for needed maintenance activities.

Community water systems shall have a professional inspection performed and a written report produced on each of their finished water and distribution storage tanks at least once every five years. Non-community water systems shall have a professional inspection and written report performed on each of their atmospheric pressure finished water and distribution storage tanks no less frequently than every five years. Records of these inspections shall be available to the Department personnel for inspection. Persons conducting underwater inspections of finished water storage tanks shall comply with AWWA Standard C652-92 or later versions of the standard.

This can be found at the following web site:
http://www.state.tn.us/sos/rules/1200/1200-05/1200-05.htm

Vermont

Water Supply Division
Dept. of Environmental Conservation
Old Pantry Building
103 South Main Street
Waterbury, VT 05671-0403
802-241-3400

Subchapter 21-7 of the Vermont Water Supply Ruler requires tanks to be inspected every 5 years
Go to www.vermontdrinkingwater.org to get a copy of the Vermont Water Supply Rule

Wisconsin

Bureau of Water Supply
Wisconsin Dept.of Natural Resources
PO Box 7921
Madison, WI 53707
Phone: 608-266-2299
Fax: 608-267-7650

Larry B Landsness, Water Supply Engineer with the Public Water Supply Section Bureau of Drinking Water & Groundwater, Wisconsin Department of Natural Resources wrote the following:

"In Wisconsin, water tanks serving public water systems are required to be emptied and inspected every 5 years per NR 811.08(5), Wis. Adm. Code. In some cases we have allowed other forms of inspections such as diving inspections or float down inspections to be substituted if the tank is new or in very good condition and not expected to need maintenance. However, we have always required someone to physically enter the tank. Enclosed are the guidelines".

General Guidelines for Underwater Inspections of Potable Water Reservoirs and Storage Tanks

There has been an increased interest in conducting underwater inspections of water storage tanks and reservoirs. This is an acceptable method of conducting routine interior inspections and cleaning in most instances. There is a section of AWWA Standard C652-02, which is the basis for the following, which is a list of requirements for underwater tank inspections in Wisconsin.

1. Underwater inspections should only be conducted on reservoirs where repairs or significant maintenance are not expected.

2. Although not mandatory, the tank should be isolated from the potable system during the inspection. The Department does not recommend on-line reservoir inspections. Unanticipated demands on the water system during on-line inspections could pose serious dangers to the divers. Also, unplanned contamination to the reservoir by the inspection team could occur and pose a health threat to the customers. The tank should remain off-line for a minimum of 15 minutes after the last diver leaves the tank.

3. A minimum free chlorine residual of 0.5 mg/l must be maintained in the tank throughout the entire inspection. Samples must be taken from the tank (not the sample tap on the riser pipe) before entering the tank and upon leaving the tank to assure the minimum chlorine residual level of 0.5 mg/l is maintained. If sediment is removed from the tank, chlorine residuals from the tank must be taken every four hours during the inspection.

4. All divers must be certified commercial divers having passed an ACDE approved course or 1st or 2nd class US Navy Diver training or equal.

5. All divers must be provided with commercial grade diving equipment.

6. All divers must use totally encapsulated diving dress including dry suit and full face sealed mask with sealed neck dam.

7. The diver's equipment must include voice communications with the surface and umbilical.

8. The inspection team must consist of a minimum of three people including at least two certified commercial divers.

9. All equipment introduced into the water must be dedicated for potable water use and must be disinfected with a minimum 200 ppm chlorine residual prior to entry.

10. The dive team must provide still photographs or color video with live voice recording to monitor all activities, findings and actions.

11. No underwater welding or coating repair is allowed.

12. A minimum of two safe bacti samples be obtained from the tank after the inspection, one following the inspection and one 24 hours later. The tank may be in service during the 24-hour period, whether the tank has been isolated or not.

13. All personnel on the dive team must be free of communicable diseases and shall not, without a physicians consent to return to diving activity, have been under a physician's care within the seven day period prior to entering the facility. No person who knowingly has an abnormal temperature or symptoms of illness shall work in a water storage facility. The water utility operator has the right to request a physician's assurance(based on an examination within the 48-hour period immediately prior to the time the diver enters the water storage facility) that all inspection personnel are free of water-transferable communicable diseases.

14. The regional DNR engineer must be informed of the date of the inspection.

States that DO NOT Require Inspections On Potable Water Storage Tanks:

Georgia

Drinking Water Permitting and Engineering Program
Georgia Environmental Protection Division
2 MLK Jr. Dr., SE., Suite 1362 East
Atlanta, GA 30334-9000
Phone: 404-651-5156; Fax: 404-651-9590

According to Drinking Water Permitting and Engineering Program Manager Onder E. Serefli, "There is no specific section, requiring regularly scheduled tank inspections, in the current Georgia drinking water rules.".

Hawaii

Environmental Management Division
Hawaii Dept. of Health
919 Ala Moana Blvd.
Room 300
Honolulu, HI 96814-4920
808-586-4258

Hawaii administrative rules do not require annual inspections or cleaning of potable water tanks.

Idaho

Division of Environmental Quality
Dept. of Health and Welfare
1410 North Hilton
Boise, ID 83706
208-373-0502

The Dept. of Health and Welfare was quick to respond to my inquiry with a phone call on 2/3/06.
A representative of the Dept. stated there are no rules for potable water tank inspections.

Kansas

Bureau of Water-Public Water Supply
Kansas Department of Health & Environment
1000 SW Jackson, Suite 420
Topeka KS 66612-1367
Phone 785 296 5503
Fax 785 296 5509

Kansas has no requirements for inspection of water storage tanks.

Kentucky

Drinking Water Branch
Division of Water
Dept. for Environmental Protection
14 Reilly Road
Frankfort Office Park
Frankfort, KY 40601
502-564-3410

According to the folks in the Drinking Water Branch there is nothing in Kentucky regulations that require inspection of ground, elevated or pressure tanks.

Michigan

Community Drinking Water Unit
Lansing Operations Division
Water Bureau, DEQ
517-241-1245

There is no statutory or mandated inspection frequency for potable storage tanks in Michigan.

According to Richard Benzie, P.E., Chief of the Community Drinking Water Unit: "The condition of storage tanks and the frequency of inspections are recorded when state staff conduct a sanitary survey of community public water systems every 3 to 5 years. If it becomes apparent that a tank or reservoir has not been inspected for too long, a recommendation will be made to do so in the near future. The acceptable inspection frequency will vary, depending upon the age of the tank, the water quality (corrosivity), the tank construction, and past inspection findings".

Nevada

No rules found. No response to our inquiry.

I tried to get each state to respond to my inquiry. However not all did. If you do not see your state listed it is due to the fact that they did not respond before we went to print.

New Mexico

Drinking Water Bureau
New Mexico Environment Department
525 Camino De Los Marquez
Suite 4
Santa Fe, NM 87501
505-827-7536

New Mexico Environment Department, Drinking Water Bureau regulations that address maintenance of storage tanks is NMAC 20.7.10.400 General Operating Requirements,

Paragraph I:

"Maintenance and disinfection of storage structures: All materials used to re-coat or repair the interior of water storage structures must be suitable for potable water contact. After the interior of a storage structure has undergone maintenance or recoating, the storage structure must be flushed and disinfected pursuant to subsection G of this Section."

A copy of our regulations can be found at: http://www.nmenv.state.nm.us/dwb/dwbtop.html

New York

Bureau of Public Water Supply Protection
Department of Health
547 River Street
Troy, NY 12180-2216
518-402-7650

There is no comparable rule in New York State that requires supplies to inspect water storage tanks on a set frequency.

Thomas N. Mottolese, Jr. of the NYS Department of Health stated; "Generally our recommendation is that storage tanks be professionally inspected every 5 to 10 years as a part of a system's routine maintenance. This is a recommendation, not a requirement".

Virginia

Division of Water Supply Engineering
Dept. of Health
Room 109-31
1500 East Main Street
Richmond, VA 23219
804-786-5566

Susan E. Douglas, P.E. Field Services Engineer for the Virginia Office of Drinking Water wrote me the following:

> Ron—I received your inquiry about inspections of potable water storage tanks. The Virginia *Waterworks Regulations* does not mandate these inspections or their frequency, as Texas does. For more information on our regulations, go to the Virginia Department of Health's website at *www.vdh.state.va.us* .

Washington

Drinking Water Division
Dept. of Health
Washington Airdustrial Center
Building 3
P.O. Box 47822
Olympia, WA 98504-7822

Jeff Johnson, P.E., Regional Engineer at the Office of Drinking Water DOH/Division of Environmental Health in Spokane, WA stated:

"We do not have a regulation that specifies an inspection frequency or inspection points, nor do we have a requirement for cleaning frequency. We do, however, have a regulation, [WAC 246-290-415(3)] that states that water systems must operate in accordance with good practices as set forth in texts, references, and manuals published by organizations such as AWWA, ASCE, etc. We do have an expectation that water systems will inspect their reservoirs on a regular basis and clean them if needed, because that is good operating practice." (sic)

Wyoming

Water Quality Division
Dept. of Environmental Quality
Herschler Building
122 West 25th Street
Cheyenne, WY 82002
307-777-5985

On 2/2/06, I received a memo from the Senior Environmental Analyst in Wyoming that stated:

Potable water storage tank inspections <u>are not required</u> in Wyoming on any schedule or time interval. Our regulations concerning "Finished Water Storage" concentrate on the proper design, materials, construction, sizing, location, protection, access, and disinfection after any construction, repair, or cleaning. If and when a finished water storage facility might need an inspection is left up to the discretion of the system owner/operator. To view our requirements concerning finished water storage in more detail go to Chapter 12 of the Wyoming DEQ Water Quality Rules and Regulations, Section 13. These can be found at: http://soswy.state.wy.us/RULES/5059.pdf

CHAPTER FOUR

Ron Perrin © 2008 Ron Perrin © 2008

Looking into the top of a potable water storage tank, less than three feet of visibility

What is in our water distribution systems?

In 1997, I started ***Ron Perrin Water Technologies*** using remote cameras to perform inspections and a potable water dive crew for cleaning potable water storage tanks. Using a underwater camera we can get a good idea of sediment depth and the condition of painted surfaces without disrupting the water system.

Sometimes we have been amazed at what we find. Things left behind from the construction of the tank such as gloves and other materials from contractors are a common occurrence. Some of the extreme cases include bats inside one tower in the hill country, and a Gila monster lizard out in west Texas.

We have inspected a few tanks that had wooden roofs. Some have been underground tanks built long ago and covered to meet new standards. I have never seen a tank covered with a wooden roof that did not have a dead mouse or dead rat in it.

The worse by far was a small community that had been given the short water tower next to the rail road track when the rail road company switched from steam to diesel. Somewhere around the 1960's the state health department noticed the tower and told the community that they were in violation of heath standards because the tower was uncovered (no roof). So, they got together and put a wooden roof on it. In the early 90's, I inspected this tank. Out in a field alone, the rail road lines and commerce a distant memory, when I opened the hatch on top of the shingled roof I shuddered. Rats! Rats everywhere, live rats, dead rats, big rats, small rats. Rats, rats, rats!

Inspections are very important. You don't know what is in your water storage tank until you look. While rats (and lots of them) were the extreme worst, birds and insects are much more common. All tanks have vent structures so proper inspections of the vent screens are very important. The rule of thumb is: If you find a hole in the vent screen you can put your finger through, there will be insects in your tank. If you have a hole you can put your fist in you probably have dead birds in your tank.

Vent screens corrode and need to be replaced from time to time. An old screen can blow off in a storm. On a hot day, the birds & insects smell the water, enter the tank through the vent structure, and can't find their way out.

Sediment in the floor of the tank can be a habitat for bacteria, viruses, protozoa, algae, fungi, and invertebrates. Shielded from treatment chemicals the sediment can allow some contaminants to grow and become a health risk.

In their National Assessment of Tap Water Quality published in 2005 the **Environmental Working Group** (EWG) found that water suppliers across the U.S. detected 260 contaminants in water served to the public.

The list of contaminants found in drinking water is staggering, ranging from heavy metals and uranium to bacteria, protozoa, and viruses. Even treatment chemicals used for disinfection can form harmful chemical by-products in the treated water.

Tap water in 42 states is contaminated with more than 140 unregulated chemicals that lack safety standards, according to the EWG's two-and-a-half year investigation of water suppliers' tests of the treated tap water served to communities across the country. In an analysis of more than 22 million tap water quality tests, most of which were required under the federal Safe Drinking Water Act, the EWG found that water suppliers across the U.S. detected 260 contaminants in water served to the public. 141 of these detected chemicals — more than half — are unregulated; public health officials have not set safety standards for these chemicals, even though millions drink them every day.

EWG's analysis also found over 90 percent compliance with enforceable health standards on the part of the nation's water utilities, showing a clear commitment to comply with safety standards once they are developed. The problem, however, is EPA's failure to establish enforceable health standards and monitoring requirements for scores of widespread tap water contaminants. Of the 260 contaminants detected in tap water from 42 states, only 114 have EPA set enforceable health limits (called Maximum Contaminant Levels, or MCLs), and for 5 others the Agency has set non-enforceable goals called secondary standards. (EPA 2005a). The 141 remaining chemicals without health-based limits contaminate water served to 195,257,000 people in 22,614 communities in 42 states.

EWG acquired tap water testing data from state water offices, which collect it from drinking water utilities to fulfill their role as primary enforcement agents. EPA does not maintain a comprehensive, national tap water quality database. Instead, the Agency sets safety standards for contaminants based on partial information, from test data it gathers from select, representative states and water suppliers. EWG will be making its data available to the EPA, state authorities and water utilities.

The statistics reported here represent an underestimate of the scope of consumers' exposure to unregulated contaminants in the nation's tap water. The state records we have compiled contain no tests whatsoever on unregulated contaminants for fully 23% of the 39,751 water systems represented, and EPA has required testing, in limited surveillance programs, for just a fraction of the hundreds of unregulated tap water contaminants identified in peer-reviewed studies. Some unregulated contaminants were found in the tap water of hundreds of communities, while others were found in very few; some were detected at levels of health concern, while others were not. These differences in the scale and magnitude of exposures can guide priorities when EPA assesses potential mandatory safety standards for these chemicals:

- Of the 141 unregulated contaminants found in tap water, 40 were detected in tap water served to at least one million people. while 20 unregulated contaminants were detected in just one system, only one time.
- Nineteen unregulated contaminants were detected above health-based limits (EPA 2004b) in tap water served to at least 10,000 people. Forty-eight unregulated contaminants were not detected above health-based limits anywhere, and seventy lack health-based limits, which have yet to be developed by EPA.

The Agency has fallen short in efforts both to require the testing that would reveal what pollutants are in tap water supplies and to set health-based standards for those that are found. EPA has ignored three mandatory Safe Drinking Water Act deadlines to set standards for unregulated contaminants (EPA 2001a). Nearly twenty percent of the contaminants that EPA is currently considering for regulation have been under study at the Agency for 17 years now, beginning with testing programs initiated in 1988 (EPA 2001b, 2005b).

The agency has also failed to act on its own information showing that increased testing is justified. EPA has required water suppliers to test tap water for approximately 200 unregulated contaminants over the past 30 years (EPA 2001b, 2001c, 2005c, FR 1996—details). But the Agency's own scientists have identified 600 chemicals in tap water formed as by-products of disinfection (Richardson 1998, 1999a,b, 2003), tracked some 220 million pounds of 650 industrial chemicals discharged to rivers and streams each year (EPA 2003), and spearheaded research on emerging contaminants after the U.S. Geological Survey found 82 unregulated pharmaceuticals and personal care product chemicals in rivers and streams across the country that provide drinking water for millions of Americans (Kolpin et al. 2004, EPA 2005d). All told, EPA has set safety standards for fewer than 20 percent of the many hundreds of chemicals that it has identified in tap water.

In addition: as recently as March 2008, (in the course of a five-month inquiry), the **Associated Press** discovered that drugs have been detected in the drinking water supplies of 24 major metropolitan areas.

The New York Sun reported in an article entitled **City Lawmakers Find 'Alarming' Report of Drugs in Water**: "Following requests for the comment regarding the report by the Associated Press, the New York City Department of Environmental Protection released a statement yesterday afternoon acknowledging concerns about the issue: 'Though nothing in the information we've seen presents a risk to this water supply, we understand and take very seriously public concerns about pharmaceuticals in drinking water and continue to closely monitor this emerging national issue, in cooperation with the New York City Department of Health and Mental Hygiene. DEP and DOHMH are working together to develop an education program about the best disposal methods for medications, targeting both watershed communities and city residents, and to consider appropriate next steps.' "

The report points out that while the quantity of pharmaceuticals found in the watersheds is minute and has not been proven to be harmful to humans, a study found that feminized male flounder in Jamaica Bay were likely affected by discharged pharmaceuticals containing forms of estrogen.

Removing sediment from your water storage tank is an effective way to ensure that ALL contaminants that have accumulated over time have been removed.

In 1990, the EPA Science Advisory Board (SAB) cited drinking water contamination as one of the most important environmental risks and indicated that disease causing microbial contaminants (i.e., bacteria, protozoa, and viruses) are probably the greatest remaining health risk management challenge for drinking water suppliers (USEPA/SAB, 1990).

EPA has determined that the presence of microbiological pathogens in public water supplies is a health concern. If finished water supplies contain microbiological contaminants, illnesses and disease outbreaks may result. Twelve waterborne cryptosporidiosis outbreaks caused by contamination in public water systems were reported to the Center for Disease Control and Prevention between 1984 and 1998. In 1993, *Cryptosporidium* caused more than 400,000 people in Milwaukee, WI, to experience intestinal illness. More than 4,000 were hospitalized and at least 50 deaths were attributed to this cryptosporidiosis outbreak. Other recent cryptosporidiosis outbreaks attributable to public water system contamination occurred in Nevada, Oregon, and Georgia.

The EPA set enforceable drinking water treatment technique requirements to reduce the risk of *Cryptosporidium* from surface water for systems serving at least 10,000 persons.

The Safe Drinking Water Act (SDWA) requires EPA to set enforceable standards to protect public health from contaminants that may occur in drinking water. As explained in more detail in the April 10, 2000 proposal for today's rule (65 FR 19046), EPA has determined that the presence of microbiological contaminants is a substantial health concern. If finished water supplies contain microbiological contaminants, disease outbreaks may result. Disease symptoms may include diarrhea, cramps, nausea, jaundice, headaches, and fatigue. EPA has set enforceable drinking water treatment techniques to reduce the risk of waterborne disease outbreaks. Treatment technologies such as filtration and disinfection can remove or inactivate microbiological contaminants.

Physical removal is critical to the control of Cryptosporidium because it is highly resistant to standard disinfection practices. Cryptosporidiosis, the infection caused by Cryptosporidium, may manifest itself as a severe infection that can last several weeks and may cause the death of individuals with compromised immune systems.

Final Long Term 1 Enhanced Surface Water Treatment Rule (LT1ESWTR)

The LT1ESWTR extends further this necessary protection from *Cryptosporidium* to communities of fewer than 10,000 persons. Today's rule for the first time establishes *Cryptosporidium* control requirements for systems serving less than 10,000 persons by requiring a minimum 2-log removal for *Cryptosporidium*. The rule also strengthens filter performance requirements to ensure 2-log *Cryptosporidium* removal, establishes individual filter monitoring to minimize poor performance in individual units, includes *Cryptosporidium* in the definition of GWUDI, and explicitly considers unfiltered system watershed control provisions.

Twelve waterborne cryptosporidiosis outbreaks have occurred at drinking water systems since 1984 (Craun, 1998; USEPA, 2000a). The largest of the known outbreaks occurred in Milwaukee and was responsible for over 400,000 illnesses and at least 50 deaths (MacKenzie et al., 1994; Hoxie, et al., 1997); other known outbreaks have occurred in smaller communities and have involved many fewer people. An incident such as a rainstorm that flushes many oocysts into the source water or causes a sanitary sewer overflow combined with a water treatment plant upset could allow a large pulse of oocysts to move past the multiple barriers of a water treatment plant.

To read more about the *"Final Long Term 1 Enhanced Surface Water Treatment Rule"*

See the Fact Sheet at: http://www.epa.gov/safewater/mdbp/lt1eswtr_fact.html or

For general information on the LT1ESWTR, contact the Safe Drinking Water Hotline, at (800) 426-4791, or visit the EPA Safewater website, www.epa.gov/safewater/mdbp/lt1eswtr.html.

For copies of the *Federal Register* notice of the final regulation or technical fact sheets, contact the Safe Drinking Water Hotline at (800) 426-4791. The Safe Drinking Water Hotline is open Monday through Friday, excluding Federal holidays, from 9:00 a.m. to 5:30 p.m. Eastern Time.

(Editors note: Having contracted and suffered with Cryptosporidium myself, I can attest to the debilitating affects of this health problem. I would urge all readers to be aware and on guard.)

RPWT © 2007 Contaminated sediment being removed from potable water tower.

If thousands of bacteria can not be seen in a glass of water, how many does it take to make the water turn BLACK?

Health Risk From Microbial Growth and Biofilms in Drinking Water Distribution Systems.

On June 17, 2002, the EPA Office of Ground and Drinking Water issued a paper on distribution systems titled "Health Risk From Microbial Growth and Biofilms in Drinking Water Distribution Systems".

In *"Section G Sediment Accumulation"* the paper explains that microbial activity may occur in accumulated sediment. Microbes have been identified in accumulated sediments, including both pathogens and non-pathogens. These include bacteria, viruses, protozoa, algae, fungi, and invertebrates. Opportunistic pathogens have also been detected in sediments. These pathogens include **Legionella** and micro-bacteria. Hepatitis A is a primary pathogen that has been documented to survive more than four months in sediment. Other pathogens you could encounter in your sediment are *Pseudomonas fluorescence* and *Flavobacterium*.

The paper covers the importance of inspecting and cleaning water storage tanks in *"section I Proper Storage Tank Vessel Management and Alteration"*. Sites Cleaning, Inspection & disinfection as one measure to bring systems back into compliance.

See the entire report at: http://www.epa.gov/safewater/disinfection/tcr/pdfs/whitepaper_tcr_biofilms.pdf

Finished Water Storage Facilities

On August 15, 2002, the EPA Office of Ground and Drinking Water issued a paper on distribution systems titled **"Finished Water Storage Facilities"**.

The paper covers the importance of inspecting water storage tanks in section 3.3. Section 3.4 points out the importance of tank cleaning. Again in the summery the report recaps the importance of inspection and cleaning to deal with micro-biological, chemical, and physical water quality problems can occur in finished water. In Section "3.4 Maintenance Activities" The paper states how Commercial diving contractors can be used to clean and inspect storage facilities that cannot be removed from service. I would add that a diving contractor can not only keep your tank in service, they can also remove ALL loose sediment in much less time than traditional methods. The paper sites AWWA Standard C652-92 for guidelines on disinfection of all equipment used to clean storage facilities. They also offer a case study of an underwater cleaning of three steel elevated spheroids at the City of Brookfield Water Utility in Brookfield, Wisconsin. "The tank with the longest cleaning interval contained the most accumulated sediment (28 inches maximum depth compared to 4-12 inches in the other two tanks), and the highest HPC bacteria levels before cleaning

(1300/mL compared to 640 and 80/mL in the other two tanks). As a result of underwater cleaning, HPC bacteria and turbidity levels were significantly reduced".

Those sections have been reprinted for your convenience. The entire paper can be seen at: http://www.epa.gov/safewater/disinfection/tcr/pdfs/whitepaper_tcr_storage.pdf

3.3 Reprint from "Finished Water Storage Facilities".

Tank Inspections

Like water quality monitoring, tank inspections provide information used to identify and evaluate current and potential water quality problems. Both interior and exterior inspections are employed to assure the tank's physical integrity, security, and high water quality. Inspection type and frequency are driven by many factors specific to each storage facility, including its type (i.e. standpipe, ground tank, etc), vandalism potential, age, condition, cleaning program or maintenance history, water quality history, funding, staffing, and other utility criteria. AWWA

Manual M42, Steel Water Storage Tanks (1998) provides information regarding inspection during tank construction and periodic operator inspection of existing steel tanks. Specific guidance on the inspection of concrete tanks was not found in the literature. However, the former AWWA Standard D101 document may be used as a guide to inspect all appurtenances on concrete tanks. Concrete condition assessments should be performed with guidance from the tank manufacturer. Soft, low alkalinity, low pH waters may dissolve the cementitious materials in a concrete reservoir causing a rough surface and exposing the sand and gravel. The concern is that in extreme cases, the integrity of reinforcing bars may be compromised. Sand may collect on the bottom of the storage facility during this process.

Routine inspections typically monitor the exterior of the storage facility and grounds for evidence of intrusion, vandalism, coating failures, security, and operational readiness. Based on a literature review and project survey, Kirmeyer et al. (1999) suggested that routine inspections be conducted on a daily to weekly basis. Where SCADA systems include electronic surveillance systems, alarm conditions may substitute for physical inspection. Periodic inspections are designed to review areas of the storage facility not normally accessible from the ground and hence not evaluated by the routine inspections. **These inspections usually require climbing the tank.** Periodic inspections, like routine inspections, are principally a visual inspection of tank integrity and operational readiness. Based on a literature review and project survey, Kirmeyer et al. (1999) suggested that periodic inspections be conducted every 1 to 4 months. **Comprehensive inspections are performed to evaluate the current condition of storage facility**

components. These inspections often require the facility to be removed from service and drained unless robotic devices or divers are used.

The need for comprehensive inspections is generally recognized by the water industry. AWWA Manual M42 (1998) recommends that tanks be drained and inspected at least once every 3 years or as required by state regulatory agencies. Most states do not recommend inspection frequencies thereby leaving it to the discretion of the utility. States that do have recommendations are Alabama (5 years), Arkansas (2 years), Missouri (5 years), New Hampshire (5 years), Ohio (5 years), Rhode Island (external once per year; internal, every five years), Texas (annually), and Wisconsin (5 years). Kirmeyer et al. (1999) recommend that comprehensive inspections be conducted every 3 to 5 years for structural condition and possibly more often for water quality purposes. Uncovered finished water reservoirs have unique problems. Consequently, water utilities have ceased constructing such facilities. As noted previously, the IESWTR prohibits construction of new uncovered finished water reservoirs in the U.S. Under the LT2ESWTR, existing uncovered finished water reservoirs will be managed in accordance with a state approved plan, if the facility is not covered subsequent to the rule's implementation. Flexible membrane covers are one means of enclosing uncovered reservoirs and these types of facilities also require specific routine, periodic, and comprehensive inspections to ensure the cover's integrity.

3.4 Maintenance Activities

Storage facility maintenance activities include cleaning, painting, and repair to structures to maintain serviceability. Based on a utility survey conducted by Kirmeyer et al. (1999), it appears that most utilities that have regular tank cleaning programs employ a cleaning interval of 2 to 5 years. This survey also showed that most tanks are painted (exterior coating) on an interval of 10 to 15 years. The following existing standards are relevant to disinfection procedures and approval of coatings: ANSI/NSF Standard 61, and Ten States Standards (Great Lakes . . . 1997) AWWA Manuals Prepared by AWWA with assistance from Economic and Engineering Services, Inc.

AWWA M25 – Flexible-Membrane Covers and Linings for Potable-Water
Reservoirs (1996) AWWA M42 – Steel Water-Storage Tanks (1998) AWWA Standards AWWA Standard C652-92 Disinfection of Storage Facilities (AWWA 1992) provides guidance for disinfection when returning a storage facility to service. AWWA Standard D102 recognizes general types of interior coating systems including: Epoxy, Vinyl, Enamel, and Coal-Tar. Each of the coating systems listed under AWWA Standard D102 has provided satisfactory service when correctly applied (AWWA 1998). Other coating systems have been successfully used including chlorinated rubber, plural-component urethanes, and metalizing with anodic material (AWWA 1998). Epoxy and solvent-less

polyurethanes interior coating systems are most likely to meet strict environmental guidelines and AWWA and NSF Standards (Jacobs 2000).

Spray metalizing using zinc, aluminum or a combination of both is also a promising alternative. Coal tar coating systems are not common in eastern U.S. as the coatings installed in the 1950s and 1960s have mostly been replaced or the tanks themselves have been removed from service. Coal tar is still in use in California where it is often applied over an epoxy system on tank floors (Lund, 2002). ANSI/NSF 61 (National Sanitation Foundation 1996) is a nationally accepted standard that protects stored water from contamination via products which come into contact with water. Products covered by NSF 61 include pipes and piping appurtenances, nonmetallic potable water materials, coatings, joining and sealing materials (i.e. gaskets, adhesives, lubricants), mechanical devices (i.e. water meters, valves, filters), and mechanical plumbing devices. NSF 61 was reviewed and certified by the American National Institute of Standards (ANSI) which permitted the use of the standard by other independent testing agencies such as Underwriters Laboratories. With the development of this ANSI/NSF-61 Standard, the approval and reporting for tank coatings process is now standardized. State agencies that previously had independent coating approval programs discontinued these programs and adopted the ANSI/NSF 61 Standard. Details on the ANSI/NSF 61 certification procedure are provided in the Permeation and Leaching White Paper. Coating manufacturers provide technical specifications for proper coating application and curing.

Utilities or their consulting engineer provide technical specifications and drawings describing the specific project. Trained and certified coating inspectors provide quality control during coating application. The National Association of Corrosion Engineers has a certification program for coating inspectors. **Kirmeyer et al. (1999) recommended that covered facilities be cleaned every three to five years, or more often based on inspections and water quality monitoring, and that uncovered storage facilities be cleaned once or twice per year. Commercial diving contractors `can be used to clean and inspect storage facilities that cannot be removed from service. AWWA Standard C652-92 provides guidelines for disinfection of all equipment used to clean storage facilities**.

As previously mentioned, the three finished water steel elevated spheroids at the City of Brookfield Water Utility in Brookfield, Wisconsin that were the subject of a field study (Kirmeyer et al. 1999) conducted to document the underwater cleaning process and its water quality impacts stated that the time since last cleaning was 15 years for one tank and 7 years for the other two tanks. The tank with the longest cleaning interval contained the most accumulated sediment (28 inches maximum depth compared to 4-12 inches in the other two tanks), and the highest HPC bacteria levels before cleaning (1300/mL compared to 640 and 80/mL in the other two tanks). As a result of underwater cleaning, HPC bacteria and turbidity levels were significantly reduced. Maintenance of the cathodic protection system is a component of controlling corrosion and degradation of the submerged coated surface of finished water storage facilities. AWWA Standard D104 (AWWA 1991) provides guidelines on system inspection and maintenance.

3.5 Operations Activities

As previously noted, water age is an **important** variable in managing water quality in finished water storage. Operationally, water age in these facilities is managed by routine turn over of the stored water and fluctuation of the water levels in storage facilities. Kirmeyer et al. (1999) recommended a 3 to 5 day complete water turnover as a starting point, but cautioned that each storage facility be evaluated individually and given its own turnover goal. Water storage management for water quality must take into account influent water quality, environmental conditions, retention of fire flow, and demand management, as well as factors specific to the design and operation of the tank such as velocity of influent water, operational level changes, and tank design. Consequently, water level fluctuations in a distribution system are managed as an integrated operation within pressure zones, demand service areas, and the system as a whole rather than on an individual tank basis.

Available guidelines for water turnover rates are summarized in Table 2. From a field perspective, the Philadelphia Water Department estimated mean residence time and turnover rate in several standpipes by measuring fluoride residual and water levels. Mean residence time of water in the standpipes was determined to be 50 percent longer than expected because "old" water re-entered the standpipes from the distribution system. One major conclusion from this work was that for water to get out into the distribution system and away from the standpipes, standpipe drawdown needs to correspond to peak demands or precede peak demands. (Burlingame, Korntreger and Lahann 1995). Philadelphia also demonstrated how operational changes can reduce the hydraulic detention time needed to restore or maintain a disinfectant residual within the storage facility. During normal operation, the water levels in the storage facilities were allowed to drop an additional ten feet in elevation, decreasing the mean residence time by two to three days. As a result, disinfectant residuals were maintained at acceptable levels, even during the summer months (Burlingame and Brock 1985).

Design of Storage Facilities

The sizing, number, and type of storage facilities affect a water system's ability to manage water quality while providing an adequate water supply with adequate pressure. Capital planning necessitates installation of facilities that have excess capacity for water storage and distribution. Standard design guidelines for hydraulic considerations in the planning and construction of tanks are available in: •AWWA Manual M32 Distribution Network Analysis for Water Utilities (AWWA 1989) •Modeling, Analysis and Design of Water Distribution Systems (AWWA 1995c) •Hydraulic Design of Water Distribution Storage Tanks (Walski 2000) These guidelines ensure adequate fire flow to meet applicable codes and rating systems as well as hydraulics of water storage. State regulations address design features related to tank sizing, siting, penetrations, coatings and linings through reference to industry recognized codes and manuals (i.e. AWWA, NSF International and 10 States Standards). A discussion relating fire flow requirements to storage volume and water age is provided in the Water Age White Paper. Findings suggest that volumetric increases are site-specific and cannot be generalized. Design guidelines addressing water quality include:

•*Maintaining Water Quality in Finished Water Storage Facilities* (Kirmeyer et al. 1999)

•*Water Quality Modeling of Distribution System Storage Facilities* (Grayman et al.2000)

Appurtenances on storage facilities, such as vents, hatches, drains, wash out piping, sampling taps, overflows, valves, catwalk, etc., can be critical to maintaining water quality. The Ten State Standards (Great Lakes . . . 1997) provides recommended design practices for appurtenances. Design considerations include mixing to preclude dead zones and to maintain a disinfectant residual. Guidelines for momentum-based mixing can be found in Grayman et al. (2000). Other types of mixings systems are described in Kirmeyer et al. (1999).

4.0 Summary

Microbiological, chemical, and physical water quality problems can occur in finished water reservoirs that are under-utilized or poorly mixed. Poor mixing can be a result of design and/or operational practices. Several guidance manuals have been developed to address design, operations, and maintenance of finished water reservoirs. Water quality issues that have the potential for impacting public health include DBP formation, nitrification, pathogen contamination, and increases in VOC/SOC concentrations. Elevated DBP levels within storage facilities could result in an MCL violation under the proposed Stage 2 Disinfectants and Disinfection Byproduct Rule, based on a locational running annual average approach. A separate White Paper on Nitrification indicates that nitrite and/or nitrate levels are unlikely to approach MCL concentrations within the distribution system due to nitrification unless finished water nitrate/nitrite levels are near their respective MCLs. Pathogen contamination from floating covers or unprotected hatches is possible. Recommended tank cleaning and inspection procedures have been developed by AWWA and AWWARF to address these issues. Elevated levels of VOCs and SOCs have been measured in finished water storage facilities. AWWA and NSF standards have been developed to ensure that approved coatings function as intended. Addition data and evaluation would be required to determine if there is a significant potential for coatings and other products used in distribution system construction and maintenance to cause an MCL violation based on sampling within the distribution system rather than the currently required monitoring at the point of entry.

See the entire report at:
http://www.epa.gov/safewater/disinfection/tcr/pdfs/whitepaper_tcr_storage.pdf

Sediment being removed from a ground tank by dive crew RPWT © 2007

Cleaning sediment is an effective way of keeping your distribution system free of contaminates.

Drinking Water Contaminants from EPA

http://www.epa.gov/safewater/contaminants/index.html#micro

List of Contaminants & their Maximum Contaminant Level (MCLs)

Microorganisms

Contaminant	MCLG[1] (mg/L)[2]	MCL or TT[1] (mg/L)[2]	Potential Health Effects from Ingestion of Water	Sources of Contaminant in Drinking Water
Cryptosporidium (pdf file)	zero	TT [3]	Gastrointestinal illness (e.g., diarrhea, vomiting, cramps)	Human and animal fecal waste
Giardia lamblia	zero	TT[3]	Gastrointestinal illness (e.g., diarrhea, vomiting, cramps)	Human and animal fecal waste
Heterotrophic plate count	n/a	TT[3]	HPC has no health effects; it is an analytic method used to measure the variety of bacteria that are common in water. The lower the concentration of bacteria in drinking water, the better maintained the water system is.	HPC measures a range of bacteria that are naturally present in the environment
Legionella	zero	TT[3]	Legionnaire's Disease, a type of pneumonia	Found naturally in water; multiplies in heating systems

Total Coliforms (including fecal coliform and E. Coli)	zero	5.0%[4]	Not a health threat in itself; it is used to indicate whether other potentially harmful bacteria may be present[5]	Coliforms are naturally present in the environment; as well as feces; fecal coliforms and E. coli only come from human and animal fecal waste.
Turbidity	n/a	TT[3]	Turbidity is a measure of the cloudiness of water. It is used to indicate water quality and filtration effectiveness (e.g., whether disease-causing organisms are present). Higher turbidity levels are often associated with higher levels of disease-causing microorganisms such as viruses, parasites and some bacteria. These organisms can cause symptoms such as nausea, cramps, diarrhea, and associated headaches.	Soil runoff
Viruses (enteric)	zero	TT[3]	Gastrointestinal illness (e.g., diarrhea, vomiting, cramps)	Human and animal fecal waste

Disinfection Byproducts

Contaminant	MCLG[1] (mg/L)[2]	MCL or TT[1] (mg/L)[2]	Potential Health Effects from Ingestion of Water	Sources of Contaminant in Drinking Water
Bromate	zero	0.010	Increased risk of cancer	Byproduct of drinking water disinfection
Chlorite	0.8	1.0	Anemia; infants & young children: nervous system effects	Byproduct of drinking water disinfection
Haloacetic acids (HAA5)	n/a[6]	0.060[7]	Increased risk of cancer	Byproduct of drinking water disinfection
Total Trihalomethanes (TTHMs)	n/a[6]	0.080[7]	Liver, kidney or central nervous system problems; increased risk of cancer	Byproduct of drinking water disinfection

Disinfectants

Contaminant	MRDLG[1] (mg/L)[2]	MRDL[1] (mg/L)[2]	Potential Health Effects from Ingestion of Water	Sources of Contaminant in Drinking Water
Chloramines (as Cl_2)	MRDLG=4[1]	MRDL=4.0[1]	Eye/nose irritation; stomach discomfort, anemia	Water additive used to control microbes
Chlorine (as Cl_2)	MRDLG=4[1]	MRDL=4.0[1]	Eye/nose irritation; stomach discomfort	Water additive used to control microbes
Chlorine dioxide (as ClO_2)	MRDLG=0.8[1]	MRDL=0.8[1]	Anemia; infants & young children: nervous system effects	Water additive used to control microbes

Inorganic Chemicals

Contaminant	MCLG[1] (mg/L)[2]	MCL or TT[1] (mg/L)[2]	Potential Health Effects from Ingestion of Water	Sources of Contaminant in Drinking Water
Antimony	0.006	0.006	Increase in blood cholesterol; decrease in blood sugar	Discharge from petroleum refineries; fire retardants; ceramics; electronics; solder
Arsenic	0[7]	0.010 as of 01/23/06	Skin damage or problems with circulatory systems, and may have increased risk of getting cancer	Erosion of natural deposits; runoff from orchards, runoff from glass & electronics—production wastes
Asbestos (fiber >10 micrometers)	7 million fibers per liter	7 MFL	Increased risk of developing benign intestinal polyps	Decay of asbestos cement in water mains; erosion of natural deposits
Barium	2	2	Increase in blood pressure	Discharge of drilling wastes; discharge from metal refineries; erosion of natural deposits
Beryllium	0.004	0.004	Intestinal lesions	Discharge from metal refineries and coal-burning factories; discharge from electrical, aerospace, and defense industries

Cadmium	0.005	0.005	Kidney damage	Corrosion of galvanized pipes; erosion of natural deposits; discharge from metal refineries; runoff from waste batteries and paints
Chromium (total)	0.1	0.1	Allergic dermatitis	Discharge from steel and pulp mills; erosion of natural deposits
Copper	1.3	TT[g]; Action Level=1.3	Short term exposure: Gastrointestinal distress Long term exposure: Liver or kidney damage People with Wilson's Disease should consult their personal doctor if the amount of copper in their water exceeds the action level	Corrosion of household plumbing systems; erosion of natural deposits
Cyanide (as free cyanide)	0.2	0.2	Nerve damage or thyroid problems	Discharge from steel/ metal factories; discharge from plastic and fertilizer factories
Fluoride	4.0	4.0	Bone disease (pain and tenderness of the bones); Children may get mottled teeth	Water additive which promotes strong teeth; erosion of natural deposits; discharge from fertilizer and aluminum factories

Lead	zero	TT[8]; Action Level=0.015	Infants and children: Delays in physical or mental development; children could show slight deficits in attention span and learning abilities Adults: Kidney problems; high blood pressure	Corrosion of household plumbing systems; erosion of natural deposits
Mercury (inorganic)	0.002	0.002	Kidney damage	Erosion of natural deposits; discharge from refineries and factories; runoff from landfills and croplands
Nitrate (measured as Nitrogen)	10	10	Infants below the age of six months who drink water containing nitrate in excess of the MCL could become seriously ill and, if untreated, may die. Symptoms include shortness of breath and blue-baby syndrome.	Runoff from fertilizer use; leaching from septic tanks, sewage; erosion of natural deposits
Nitrite (measured as Nitrogen)	1	1	Infants below the age of six months who drink water containing nitrite in excess of the MCL could become seriously ill and, if untreated, may die. Symptoms include shortness of breath and blue-baby syndrome.	Runoff from fertilizer use; leaching from septic tanks, sewage; erosion of natural deposits

Selenium	0.05	0.05	Hair or fingernail loss; numbness in fingers or toes; circulatory problems	Discharge from petroleum refineries; erosion of natural deposits; discharge from mines
Thallium	0.0005	0.002	Hair loss; changes in blood; kidney, intestine, or liver problems	Leaching from ore-processing sites; discharge from electronics, glass, and drug factories

Organic Chemicals

Contaminant	MCLG[1] (mg/L)[2]	MCL or TT[1] (mg/L)[2]	Potential Health Effects from Ingestion of Water	Sources of Contaminant in Drinking Water
Acrylamide	zero	TT[2]	Nervous system or blood problems; increased risk of cancer	Added to water during sewage/ wastewater treatmen t
Alachlor	zero	0.002	Eye, liver, kidney or spleen problems; anemia; increased risk of cancer	Runoff from herbicide used on row crops
Atrazine	0.003	0.003	Cardiovascular system or reproductive problems	Runoff from herbicide used on row crops
Benzene	zero	0.005	Anemia; decrease in blood platelets; increased risk of cancer	Discharge from factories; leaching from gas storage tanks and landfills

Benzo(a)pyrene (PAHs)	zero	0.0002	Reproductive difficulties; increased risk of cancer	Leaching from linings of water storage tanks and distribution lines
Carbofuran	0.04	0.04	Problems with blood, nervous system, or reproductive system	Leaching of soil fumigant used on rice and alfalfa
Carbon tetrachloride	zero	0.005	Liver problems; increased risk of cancer	Discharge from chemical plants and other industrial activities
Chlordane	zero	0.002	Liver or nervous system problems; increased risk of cancer	Residue of banned termiticide
Chlorobenzene	0.1	0.1	Liver or kidney problems	Discharge from chemical and agricultural chemical factories
2,4-D	0.07	0.07	Kidney, liver, or adrenal gland problems	Runoff from herbicide used on row crops
Dalapon	0.2	0.2	Minor kidney changes	Runoff from herbicide used on rights of way
1,2-Dibromo-3-chloropropane (DBCP)	zero	0.0002	Reproductive difficulties; increased risk of cancer	Runoff/leaching from soil fumigant used on soybeans, cotton, pineapples, and orchards
o-Dichlorobenzene	0.6	0.6	Liver, kidney, or circulatory system problems	Discharge from industrial chemical factories
p-Dichlorobenzene	0.075	0.075	Anemia; liver, kidney or spleen damage; changes in blood	Discharge from industrial chemical factories

1,2-Dichloroethane	zero	0.005	Increased risk of cancer	Discharge from industrial chemical factories
1,1-Dichloroethylene	0.007	0.007	Liver problems	Discharge from industrial chemical factories
cis-1,2-Dichloroethylene	0.07	0.07	Liver problems	Discharge from industrial chemical factories
trans-1,2-Dichloroethylene	0.1	0.1	Liver problems	Discharge from industrial chemical factories
Dichloromethane	zero	0.005	Liver problems; increased risk of cancer	Discharge from drug and chemical factories
1,2-Dichloropropane	zero	0.005	Increased risk of cancer	Discharge from industrial chemical factories
Di(2-ethylhexyl) adipate	0.4	0.4	Weight loss, liver problems, or possible reproductive difficulties.	Discharge from chemical factories
Di(2-ethylhexyl) phthalate	zero	0.006	Reproductive difficulties; liver problems; increased risk of cancer	Discharge from rubber and chemical factories
Dinoseb	0.007	0.007	Reproductive difficulties	Runoff from herbicide used on soybeans and vegetables
Dioxin (2,3,7,8-TCDD)	zero	0.00000003	Reproductive difficulties; increased risk of cancer	Emissions from waste incineration and other combustion; discharge from chemical factories
Diquat	0.02	0.02	Cataracts	Runoff from herbicide use

Endothall	0.1	0.1	Stomach and intestinal problems	Runoff from herbicide use
Endrin	0.002	0.002	Liver problems	Residue of banned insecticide
Epichlorohydrin	zero	TT[2]	Increased cancer risk, and over a long period of time, stomach problems	Discharge from industrial chemical factories; an impurity of some water treatment chemicals
Ethylbenzene	0.7	0.7	Liver or kidneys problems	Discharge from petroleum refineries
Ethylene dibromide	zero	0.00005	Problems with liver, stomach, reproductive system, or kidneys; increased risk of cancer	Discharge from petroleum refineries
Glyphosate	0.7	0.7	Kidney problems; reproductive difficulties	Runoff from herbicide use
Heptachlor	zero	0.0004	Liver damage; increased risk of cancer	Residue of banned termiticide
Heptachlor epoxide	zero	0.0002	Liver damage; increased risk of cancer	Breakdown of heptachlor
Hexachlorobenzene	zero	0.001	Liver or kidney problems; reproductive difficulties; increased risk of cancer	Discharge from metal refineries and agricultural chemical factories
Hexachlorocyclopentadiene	0.05	0.05	Kidney or stomach problems	Discharge from chemical factories

Lindane	0.0002	0.0002	Liver or kidney problems	Runoff/leaching from insecticide used on cattle, lumber, gardens
Methoxychlor	0.04	0.04	Reproductive difficulties	Runoff/leaching from insecticide used on fruits, vegetables, alfalfa, livestock
Oxamyl (Vydate)	0.2	0.2	Slight nervous system effects	Runoff/leaching from insecticide used on apples, potatoes, and tomatoes
Polychlorinated biphenyls (PCBs)	zero	0.0005	Skin changes; thymus gland problems; immune deficiencies; reproductive or nervous system difficulties; increased risk of cancer	Runoff from landfills; discharge of waste chemicals
Pentachlorophenol	zero	0.001	Liver or kidney problems; increased cancer risk	Discharge from wood preserving factories
Picloram	0.5	0.5	Liver problems	Herbicide runoff
Simazine	0.004	0.004	Problems with blood	Herbicide runoff
Styrene	0.1	0.1	Liver, kidney, or circulatory system problems	Discharge from rubber and plastic factories; leaching from landfills
Tetrachloroethylene	zero	0.005	Liver problems; increased risk of cancer	Discharge from factories and dry cleaners
Toluene	1	1	Nervous system, kidney, or liver problems	Discharge from petroleum factories

Toxaphene	zero	0.003	Kidney, liver, or thyroid problems; increased risk of cancer	Runoff/leaching from insecticide used on cotton and cattle
2,4,5-TP (Silvex)	0.05	0.05	Liver problems	Residue of banned herbicide
1,2,4-Trichlorobenzene	0.07	0.07	Changes in adrenal glands	Discharge from textile finishing factories
1,1,1-Trichloroethane	0.20	0.2	Liver, nervous system, or circulatory problems	Discharge from metal degreasing sites and other factories
1,1,2-Trichloroethane	0.003	0.005	Liver, kidney, or immune system problems	Discharge from industrial chemical factories
Trichloroethylene	zero	0.005	Liver problems; increased risk of cancer	Discharge from metal degreasing sites and other factories
Vinyl chloride	zero	0.002	Increased risk of cancer	Leaching from PVC pipes; discharge from plastic factories
Xylenes (total)	10	10	Nervous system damage	Discharge from petroleum & chemical factories.

Radionuclides

Contaminant	MCLG[1] (mg/L)[2]	MCL or TT[1] (mg/L)[2]	Potential Health Effects from Ingestion of Water	Sources of Contaminant in Drinking Water
Alpha particles	none[7] — zero	15 picocuries per Liter (pCi/L)	Increased risk of cancer	Erosion of natural deposits of certain minerals that are radioactive and may emit a form of radiation known as alpha radiation
Beta particles and photon emitters	none[7] — zero	4 millirems per year	Increased risk of cancer	Decay of natural and man-made deposits of certain minerals that are radioactive and may emit forms of radiation known as photons and beta radiation
Radium 226 and Radium 228 (combined)	none[7] — zero	5 pCi/L	Increased risk of cancer	Erosion of natural deposits
Uranium	zero	30 ug/L as of 12/08/03	Increased risk of cancer, kidney toxicity	Erosion of natural deposits

December 2006 Office of Water (4601M), Office of Ground Water and Drinking Water Total Coliform Rule Issue Paper titled "Inorganic Contaminant Accumulation in Potable Water Distribution Systems"

Section 3.1 list the regulated inorganic compounds including:
   Antimony, arsenic, barium, beryllium, cadmium, chromium, copper, Cyanide, Fluoride, Lead, Mercury, Nitrite and nitrate, Selenium, and Thallium.

Section 3.2 covers unregulated inorganic compounds including:
   Aluminum, Manganese, Nickel, Silver, Vanadium, Zinc and rare-earth elements (REE) are scandium (Sc), yttrium (Y), and the 15 elements on the periodic table between and including lanthanum (La) and lutetium (Lu) including dysprosium (Dy). These elements are all trivalent and have very similar chemical properties

Section 3.3 covers regulated radiological compounds including:
   Alpha Emitters, Beta and Photon Emitters, Radium, Radon, and Uranium.

Section 4.0 gives an overview of inorganic contaminant reservoirs revisits biofilms defined in the Distribution System White Paper (USEPA, 2002) entitled *Health Risks from Microbial Growth and Biofilms in the Distribution System* as a "complex mixture of microbes, organic and inorganic material accumulated amidst a microbially-produced organic polymer matrix attached to the inner surface of the distribution system." Biofilms are present in many distribution systems and often include bacteria that can be found in source water. There are two general areas of concern associated with biofilms and inorganic accumulation/release:

Section 4.3 Covers Sediments
   "Sediments are loose particulate matter that accumulates in the low-velocity zones, dead ends, and reservoirs of distribution systems. Sediments may be comprised of silt, dirt, and organics from raw water sources, minerals such as iron and manganese, corrosion by-products from the distribution system, treatment by-products, or organic matter and cell debris which grow and detach in the distribution system. Over time, sediments may remain loose, become cohesive or compacted, or adhere to pipe surfaces. If sediments remain loose, they may travel within the distribution system due to changes in flow velocity and/or direction. This section presents a general review of sediment accumulation in water distribution systems and factors affecting these accumulations.

Composition of Sediments

Sediments occur within pipelines and storage facilities to some degree in all systems. Because of their long residence time, storage facilities have the potential to accumulate significant amounts of sediment. The composition of sediments depends on the source(s) contributing to the accumulation. The source water can provide a continuous supply of sediments, such as inorganic contaminants and organic matter attached to turbidity. This is particularly true for high-turbidity groundwater sources that have the potential to introduce water with turbidities in the range of more than 1.2 NTU to the distribution system. Some of these ground water sources may have recently been designated as Ground Water Under the Influence of Surface Water. As such, additional treatment is required, which may include practices that reduce turbidity. However, if these sources have been in operation for many years, it is possible that significant levels of sediment may have already accumulated within the distribution system."

This paper goes into great detail investigating where sediment in water systems comes from and what is in them. They even review a 1993 survey conducted in the United Kingdom compassing 71 supply zones. That survey showed that there was a large variation in the quantity of deposits in water mains within supply zones.

A wide variety of factors were found to influence sediment formation including: ***Turbidity, Flow Velocity, Oxidation/Reduction Potential, & Mechanical Disturbances.***

Sediment, Scale, and Biofilm Complex can be a problem in public water systems. "The accumulation of contaminants onto the sediment/biofilm/scale complex in the distribution system is a complicated chemical, physical, and sometimes biochemical process. These processes are also dynamic, which can result in continuous or intermittent accumulation and release of contaminants depending on system-specific and site-specific conditions. The degree to which each reservoir develops independently or as a complex is not well understood at this time. Thus, the degree to which each reservoir independently or as a tri-fold complex accumulates and/or releases inorganic contaminants is not well understood."

This is a very important publication and is currently available free online through the EPA web site. Google "Inorganic Contaminant Accumulation in Potable Water Distribution Systems". Although I covered some interesting points of this paper, I urge you to read this information in its entirety.

What I do want to emphasize is that there are many factors influencing sediment formation. This chapter started out reviewing a study that found 260 contaminants in water served to the public here in the U.S.

The importance of comprehensive inspections and a regular cleaning program can not be overstated. You do not know what is in your tanks until you look. Water testing alone is not enough. Sediment can become a habitat to hide a host of contaminates.

Your best defense is to know what you're dealing with. Hire a inspection contractor who will allow you to see what is going on without taking your tanks out of service. By using a remote underwater camera, a remotely operated vehicle (ROV), or a potable water diver you are able to not only see if you have sediment in you storage tanks you can also get a look at your water. This is often a shock, as it looks clear in an 8 ounce glass, but often when you look through a few hundred thousand gallons, the visibility is less than you would think.

A professional inspection allows you to have solid documentation of the paint condition, sediment level and even a visual evaluation of your water itself.

Photo: Ron Perrin inspecting the discharge of a tank being cleaned
RPWT © 2007

Summary-

A comprehensive inspection is a critical component in the proper maintenance of your system. AWWA Manual M42 (1998) recommends that tanks be drained and inspected *at least* once every 3 years or as required by state regulatory agencies. Most states do not recommend inspection frequencies thereby leaving it to the discretion of the utility.

The fact is you don't know what you have until you look. Robotic cameras or a potable water dive crew offer a method of getting that information without taking your tank out of service.

In an analysis of more than 22 million tap water quality tests, most of which were required under the federal Safe Drinking Water Act, *Environmental Working Group* found that water suppliers across the U.S. detected 260 contaminants in water served to the public. Sediment that builds up over time in the floor of the storage tank can become a habitat for a variety of different contaminants allowing them a safe area to grow and thrive away from the chlorine and other treatment chemicals.

Recommended tank cleaning and inspection procedures have been developed by contractors to address these issues. Facilities should be cleaned every three to five years, or more often based on inspections and water quality monitoring. You do not know what is in your tank until you look. Hire an inspection contractor with an underwater camera system to see what is there. Commercial diving contractors can be used to clean and inspect storage facilities that cannot be removed from service.

CHAPTER FIVE

Security of Water Systems

© Ron Perrin 2006

The Value of a Vulnerability Assessment
and Security Surveys

Due to the changing face of terrorism in the world, this is the chapter that is most likely to change with future editions of this book. Security of our water system is a job that requires us all working together. Managers of water systems need to understand the value of vulnerability assessments and how to use them to limit risk and where risk can not be avoided develop a working Emergency Response Plan.

Formerly licensed as a Security Consultant in the State of Texas, I have performed vulnerability assessments on many major water systems in and around Texas. Since September 11th 2001, the security of the nation's water supplies has been taken to a much higher level. As directors, managers, employees or contractors, we must all think about the security of our systems. We will review mandated requirements for systems that serve over 3,300 people in addition to practical steps that smaller systems may employ to secure their facilities.

Although 100% of larger water systems have submitted a vulnerability assessment to the EPA like many federally mandated documents on a rigid time frame, the actual usefulness of the document may be in question. If the vulnerability assessment was done by an outside engineering firm and is so complex it is not easily understood it may have not served a purpose other than to make the EPA happy.

Vulnerability assessments (VA) should be updated every five years to keep them and the Emergency Response Plans they generate up to date. If your department has made major upgrades in your security or has a VA that is a little too complicated to be of use there is another tool from the private sector.

A **SECURITY SURVEY** is a commercial/industrial version of a vulnerability Assessment. This is the tool that corporations have been using for years to evaluate the security of their facilities. This should be performed by a licensed consultant with references. This can give you a new viewpoint of your water system for much less than an engineering firm would charge for a Vulnerability Assessment, covering the same basic elements as the more expensive VA. The *Security Survey* should be used on a yearly basis to keep your facilities changing security needs up to date.

© Ron Perrin 2006

Small Water Systems May Benefit
MOST from Vulnerability Assessments

ARE YOU A TARGET?

In the United States there are 311 large water systems that serve about 43% of the total population. This was the first focus of the EPA. Vulnerability Assessments were required to be submitted to the EPA by March 31, 2003. Congress thought that it was so crucial that these systems do the Assessment that they passed a bill authorizing $200,000,000.00 in grants for larger systems to employ engineering firms and Security Consultants to perform these complex Assessments and make some security improvements. For the smaller systems serving 3,300 to 100,000 $15,000,000.00 was authorized for technical assistance that took the form of free training offered through several national associations. However if your system is smaller than 3,300 nothing was required and not even training was offered.

Although the PUBLIC HEALTH SECURITY AND BIOTERRORISM PREPAREDNESS AND RESPONSE ACT OF 2002 does not require systems serving fewer than 3300 customers to perform Vulnerability Assessments on their systems many in the water and security industry believe these assessments may be a asset to any size system. The vulnerability assessments were required to include, but not be limited to, a review of pipes and constructed conveyances, physical barriers, water collection, pretreatment, treatment, storage and distribution facilities, electronic, computer or other automated systems which are utilized by the public water system, the use, storage, or handling of various chemicals, and the operation and maintenance of such system.

Although I think it is important that any Security Consultant be certified and experienced to perform a full Vulnerability Assessment to meet EPA standards, that type of report would be overkill for a yearly inspection. I also feel that a team approach is worlds better than hiring an individual. All systems are complex when it comes to security issues and a trained team will just do a more thorough job that a single person. See the appendix section to meet our Security Consulting Team or visit www.ronperrin.com.

In September 2004 I attended the Water Security Congress in Charlotte NC. I was able to meet several key players in the war on Terrorism including Chris Swecker. Chris is a Special Agent in charge of Federal Bureau of Investigations operations in the Charlotte, NC area. Mr. Swecker was stationed in Iraq in the last year, heading the FBI effort.

I was also able to visit with Robert J. Walters who was the 2004 Chair of The North Carolina American Water Works Association and Water environment Association. When asked about the benefit that smaller water utilities would gain from doing a Vulnerability Assessment, Robert stated "I think that all water utilities would gain a lot of insight, and be better prepared for all emergencies, and disasters,(i.e. ice storms, hurricanes, tornados, wind storms, floods, vandalism, leaks, equipment failures, and terrorism).

I was also able to meet Janet Pawlukiewicz. Janet was the Acting Director of the Water Security Division of the EPA. Pawlukiewicz addressed the luncheon on the second day of the conference and reported that the EPA had gotten 100% of the large city Vulnerability Assessments, and 90% of the medium sized city V.A's. EPA provides grants to organizations that provide training, technical assistance, and tool development for water security. Beginning in 2002, for example, EPA provided counterterrorism grants to ensure that drinking water utilities receive technical assistance and training on homeland security issues, including vulnerability assessments and emergency response plans. During one of the training classes Ms. Pawlukiewicz announced to the class that the EPA had just granted two million dollars to the Water ISAC to allow systems serving fewer than 3300 customers to be a member of the water ISAC. This is great news for smaller systems. I will take some time to explain what the ISAC is and why you would want to use it no matter what the size of your system is.

The Water ISAC

ISAC stands for **I**nformation **S**haring and **A**nalysis **C**enter. The Water ISAC was developed to provide America's drinking water and wastewater systems with a secure Web-Based environment for early warning of potential physical, contamination, and cyber threats and a source of knowledge about security. The Water ISAC was officially launched in December 2002. This highly secure Internet portal is the best source for sensitive security information and alerts to help America's drinking water and wastewater community protect consumers and the environment.

It is the only centralized resource that gathers, analyzes and disseminates threat information that is specific to the water community. The Water ISAC enables fast and cost-effective access to sensitive information about cyber, physical, and contamination issues.

Emergency Response Plans

Emergency response plans describe the actions that a drinking water or wastewater utility would take in response to a major event, such as natural disasters or man-made emergencies. They should address the issues raised by the utility's <u>Vulnerability Assessment</u>. The information given below is of vital importance to water utilities, but it is also a valuable resource for other government and private sector entities, such as public health and law enforcement officials, emergency responders, laboratories, technical assistance providers, guidance documents and planning tools to improve the quality and efficiency of measures taken to respond to accidental or purposeful contamination incidents.

Utilities are strongly encouraged to regularly review and update their vulnerability assessments and emergency response plans due to changes in our systems and the communities we serve.

I.   **Before You Begin Developing or Revising Your ERP:** Describes steps and actions you would need to complete before you could successfully develop or revise your ERP.

II. **Emergency Response Plan—Eight Core Elements:** Describes core elements that are universal to any ERP. If you are beginning to develop your ERP, you should use this section as a general template. If you have an existing ERP, you can use this section to check if your existing ERP is comprehensive and complete. The core ERP elements are:

   ❑ System Specific Information

   ❑ CWS Roles and Responsibilities

   ❑ Communication Procedures: Who, What, and When

   ❑ Personnel Safety

   ❑ Identification of Alternate Water Sources

   ❑ Replacement Equipment and Chemical Supplies

   ❑ Property Protection

   ❑ Water Sampling and Monitoring

III. **Putting Your ERP Together and ERP Activation:** Describes steps and issues you need to address once you have all your core ERP elements in place and how to put these elements together into a single comprehensive plan. Additionally, you will need to understand the types of events that will trigger use of the plan. An effective ERP now needs to address intentional acts of terrorism as well as other emergencies and natural disasters. Planning for these events makes developing, updating, and deciding to activate the

ERP more challenging than in the past.

IV.   **Action Plans:** Describes how Action Plans (AP) are developed and used to tailor emergency response actions to specific incidents or events. Under the Bioterrorism Act, you are required to address the findings of a VA in an ERP. An AP identifies the steps to take to address specific vulnerabilities and respond to a given incident.

V.   **Next Steps:** Describes steps to take after completing an ERP, for example, submitting a certification to EPA, conducting training, and updating your plan.

Will my ERP contain sensitive information?

Your ERP may contain sensitive information, so you should consider steps you need to take to ensure the security of your ERP. Sensitive information should be placed in appendices, or in sections that are not readily available to unauthorized personnel. The ERP, however, should be easily accessible to authorized personnel and should be easily identifiable during a major event. Steps taken to limit access by unauthorized persons should consider local and state Freedom of Information Act (FOIA) laws. Alternatively, you can opt to make your ERP general in nature so that everyone can use it and not include specific information about system vulnerabilities.

A secure copy of your ERP should be maintained in an off-premises location in the event that your primary copy cannot be accessed.

Your ERP may contain sensitive information, so you should consider steps you need to take to ensure the security of your ERP. Sensitive information should be placed in appendices, or in sections that are not readily available to unauthorized personnel. The ERP, however, should be easily accessible to authorized personnel and should be easily identifiable during a major event. Steps taken to limit access by unauthorized persons should consider local and state Freedom of Information Act (FOIA) laws. Alternatively, you can opt to make your ERP general in nature so that everyone can use it and not include specific information about system vulnerabilities.

A secure copy of your ERP should be maintained in an off-premises location in the event that your primary copy cannot be accessed.

Get more information on ERP's at: http://cfpub.epa.gov/safewater/watersecurity/

EPA's Security Product Guide

Recent events have created a heightened awareness of security at the nation's critical infrastructure, including its drinking water and wastewater systems. These systems are potentially vulnerable to different kinds of natural disasters and terrorist threats. EPA has developed a series of Security Product Guides to assist treatment plant operators and utility managers in reducing risks from, and providing protection against, possible natural disasters and intentional terrorist attacks.

The guides provide information on a variety of products available to enhance physical security (such as walls, gates, and manhole locks to delay unauthorized entry into buildings or pipe systems) and electronic or cyber security (such as computer firewalls and remote monitoring systems that can report on outlying processes). Other guides present information on monitoring tools that can be used to identify anomalies in process streams or finished water that may represent potential threats. Individual products evaluated in these guides will be applicable to distribution systems, wastewater collection systems, pumping stations, treatment processes, main plant and remote sites, personnel entry, chemical delivery and storage, SCADA, and control systems for water and wastewater treatment systems.

RPWT © 2008

A hard copy of the Security Product Guide published spring 2004 (EPA Document Number 817-B-04-001) may be ordered by calling EPA's National Service Center for Environmental Publications at (800) 490-9198

EPA's Homeland Security Research

EPA's Homeland Security Research is helping to protect human health and the environment from intentional acts of terror with an emphasis *on decontamination & consequence management, water infrastructure protection,* and *threat & consequence assessment.* EPA is working to develop tools and information that will help prevent and detect the introduction of contaminants into buildings or water systems, as well as decontaminate and disposal of contaminated materials should contamination occur. The focus of these efforts is aimed at providing advice, guidance and scientific expertise to emergency response personnel, decision-makers, and government officials that will result in improved protection for all citizens.

For more information see: http://www.epa.gov/ordnhsrc/

CHAPTER SIX

Inspection of Water Storage

RPWT © 2007
**Ron Perrin Water Technologies inspector using a
remote camera to inspect a water tower in West Texas.**

As we read in chapter three, many states require some sort of inspection. Some states, including Texas, require each of the system's ground and elevated tanks to be inspected annually, and pressure tanks once every five years by water system personnel or a contracted inspection service. What exactly is required varies from state to state. For some, climbing to the top of the facility and insuring the vent screen is in place, the water level indicator is working, and the top hatch is properly locked is a complete inspection.

BUT HAVE YOU REALLY DONE AN INSPECTION?

Many managers make the mistake of sending system personnel out to inspect water storage tanks with clip board and check list. While this may be an appropriate inspection for a water storage tank holding 10,000 gallons or smaller there are many factors larger facilities have and need a closer look and possibly a contracted inspection service.

What are the factors that would lead you to a Contract Inspector?

Water visibility—Can the inspector see the floor of the tank? Is the tank small enough that they can look through the water and see how deep sediment is, what kind of sediment is on the floor and possible contaminates. In a tank that is 10,000 gallons or less you may be able to see the floor of the tank, but for larger tanks the water may need to be removed or underwater cameras be used to see the floor. Contaminants come in many forms and shapes, and can range from bacteria to insects and birds.

These contaminants can enter your system from a wide range of entry points. We have seen new facilities with large amounts of sediment due to dirt being left in the water lines during construction then being pumped into the tank on the very first day of use. There have also been small animals make their way into water systems in this same way. Other more common entry points are the hatch being left open or a missing vent screen that could have been blown off in a storm or simply rotted way from corrosion.

Although it is currently a common practice to inspect water tank's with nothing more than a clipboard and flashlight, if you can't see the bottom of the tank you do not know what is there. In Texas for example, the state doesn't require the tank be drained to be inspected but they do want the sediment level documented. Therefore to do the inspection properly the water should be drained from the tank until the inspector can see the floor of the tank and observe if there is sediment and if there are any other contaminants present. To avoid draining the tank, inspection contractors are often hired that can deploy divers or specially constructed underwater cameras that can enter the system while it is still in service and properly inspect the interior conditions.

Personnel Safety—We are talking about the height of the tank or tower. Water tower's and standpipes are often over 150 feet tall. Even ground tanks can be 40 to 60 feet tall with straight ladders. Many water storage tanks were built before the current safety climb rules and have not yet been retrofitted with safety climbing devices, platforms, or ladders on water storage tanks where required.

Water utility personnel seldom have the proper training and experience to climb such tall ladders. In fact, we have seen the Risk Management Officer for many major cities' refuse to let any city employee climb ladders over 8 ft tall. An experienced contractor will be able to provide inspectors who have training and experience to safely climb your storage tank. In addition to having the proper gear, these people make their living climbing on tanks and towers.

Documentation—There are four major reasons to hire a contract inspection service to properly document the condition of your water storage tank inside and out:

1. **State Requirements—See chapter Three**

2. **Water Quality**—good documentation of your water

    Quality is essential. Photos and video of the water condition and sediment found inside the tank are proof that you are doing everything you can to protect the public. If you find the tank has heavy sediment that could support the growth of bacteria, the tank can be removed from service and cleaned out before it becomes a public health concern.

3. **Structural soundness**—Is the protective coating inside the paint still doing its job, or has the chlorine gases and humidity taken their toll on the painted surfaces above the waterline inside of the tank. Photos and/or video can give you a first hand look at what is going on. With this knowledge repairs can be planned out years in advance. The life of a storage tank can be greatly extended by proper and timely maintenance.

    A qualified inspection contractor will be able to show you key points on the inside roof of the tank. Documenting and keeping track of corrosion above and below the waterline allows the utility manager to plan out refurbishing and save more costly repairs.

4. **Budget expenditure justification**—A photo is worth a thousand words, regardless of the problem, from security fence repair or replacement, to tank repainting or cleaning, or other such concerns. Documentation from a qualified contractor can make getting funds from your governing body much easier. The photos and/or video of your problem areas take it from being your problem (your opinion) to being their problem, making it easier to get your job done. At the same time, if funds are refused, you have the documentation that shows you did everything in your power to do your job.

LADDER SAFETY

The current OSHA standard, 29 CFR §1910.27(d)(1)(ii), requires that safety cages or wells shall be provided on ladders of more than 20 feet to a maximum unbroken length of 30 feet.

However, the source document for 29 CFR §1910.27, ANSI A14.3-1956, Safety Code for Fixed Ladders, has been revised several times since its adoption by OSHA in 1971, and its most current version, ANSI A14.3-2002, "American National Standard for Ladders — Fixed — Safety Requirements," allows fixed ladder usage without cages or wells for a ladder length of up to 24 feet. In addition, on May 2, 2003, OSHA reopened its Proposed Rule for Walking and Working Surfaces and Personal Protective Equipment, see 68 *Federal Register* 23528. This proposed rule would amend §1910.27(d)(1)(ii) to reflect the current ANSI standard of 24 feet.

If an employer is not in compliance with the requirements of an OSHA standard, but is complying with the requirements of a proposed OSHA standard or a current consensus standard that clearly provides equal or greater employee protection, the violation of OSHA's requirement will be treated as a *de minimis* violation. *De minimis* violations are those having no direct or immediate relationship to safety and health and result in no citation, penalty, or requirement to abate.

The current OSHA standard at §1910.27(d)(2) establishes a maximum limit of 30 feet between platforms; therefore, ladder distances in excess of 30 feet without an intermittent platform would not be in compliance with the standard. However, the current ANSI A14.3-2002 standard provides that landing platforms are not required on fixed ladders with cages less than 50 feet in length; for fixed ladders with cages extending a maximum unbroken length of 50 feet, landing platforms are required at 50-foot intervals. In addition, the proposed OSHA ladder standard at §1910.23 allows for landing platforms to be installed at 50-foot intervals instead of the existing 30-foot interval for fixed ladders with cages which is mandated by the current §1910.27.
OSHA would regard compliance with the proposed requirement that landing platforms be installed at 50-foot intervals as a *de minimis* violation of the OSHA standard for fixed ladders with cages.

When ladder safety devices are used with fixed ladders on tower ladders, platforms are not required.

For changes to these guidelines consult OSHA's website at <u>http://www.osha.gov</u>. If you need further assistance, please contact the Office of General Industry Enforcement at (202) 693-1850.

CONFINED SPACE HAZARDS

Before you enter a confined water storage tank to inspect it, be aware that it is a confined space. Chemical residue on the side walls may deplete oxygen levels. On a hot summer day chlorine residue on the interior walls of a metal storage tank can turn into a deadly gas as the tank heats up.

Oxygen may also be depleted from corrosion inside the tank.

Confined spaces may be encountered in virtually any occupation; therefore, their recognition is the first step in preventing fatalities. Since deaths in confined spaces often occur because the atmosphere is oxygen deficient or toxic, confined spaces should be tested prior to entry and continually monitored. More than 60% of confined space fatalities occur among would-be rescuers; therefore, a well-designed and properly executed rescue plan is a must. There is an alert that describes 16 deaths that occurred in a variety of confined spaces. Had these spaces been properly evaluated prior to entry, continuously monitored while the work was being performed, and had appropriate rescue procedures been in effect, none of the 16 deaths would have occurred. There are no specific OSHA rules that apply to all confined spaces. Recommendations for Recognition, Testing, Evaluation, and Monitoring, and Rescue of Workers are presented. Other National Institute for Occupational Safety and Health (NIOSH) publications on this subject as well as a source for additional information and assistance are also presented.

The Regulatory Program of the United States Government (Confined Spaces [29 CFR 1910], page 282 dated August, 1985), "there are no specific OSHA rules directed toward all confined-space work, forcing OSHA compliance personnel to cite other marginally applicable standards or section 5(a)(1) in cases involving confined spaces. For this reason, OSHA field personnel have frequently and strongly recommended the promulgation of a specific standard on confined spaces." In the document ***Criteria for a Recommended Standard . . . . Working in Confined Spaces***, the National Institute for Occupational Safety and Health (NIOSH) has provided comprehensive recommendations for assuring the safety and well-being of persons required to work in confined spaces including a proposed classification system and checklist that may be applied to different types of confined spaces.

INSPECTING YOUR OWN TANK

Broken down to each inspection point, I listed a comprehensive guide to follow to perform your own tank inspection or incorporate as rules for your state or organization. Most, if not all of these requirements are based on AWWA & TCEQ guidelines.

Basics—Ground and elevated storage tank inspections should determine that the vents are in place and properly screened, the roof hatches closed and locked, flap valves and gaskets provide adequate protection against insects, rodents, and other vermin, the interior and exterior coating systems are continuing to provide adequate protection to all metal surfaces, and the tank remains in a watertight condition.

.

Ladder on a water tower in west Texas

Vents

Roof vents should be gooseneck or roof ventilator and be designed by the engineer based on the maximum outflow from the tank. Vents shall be installed in strict accordance with current AWWA standards and should be equipped with approved screens to prevent entry of animals, birds, insects and heavy air contaminants. Screens should be fabricated of corrosion-resistant material and shall be 16-mesh or finer. Screens shall be securely clamped in place with stainless or galvanized bands or wires and shall be designed to withstand winds of not less than tank design criteria (unless specified otherwise by the engineer). Avoid using stainless steel screens as they seem to corrode much faster due to the chlorine gases.

Vents on a water tower in west Texas

Roof Hatch

All roof openings should be designed in accordance with current AWWA standards. If an alternate 30 inch diameter access opening is not provided in a storage tank, the primary roof access opening should not be less than 30 inches in diameter. Other roof openings required only for ventilating purposes during cleaning, repairing, or painting operations shall be not less than 24 inches in diameter or as specified by the licensed professional engineer. An existing tank without a 30-inch in diameter access opening must be modified to meet this requirement when major repair or maintenance is performed on the tank. Each access opening shall have a raised curbing at least four inches in height with a lockable cover that overlaps the curbing at least two inches in a downward direction.

Where necessary, a gasket should be used to make a positive seal when the hatch is closed. All hatches should remain locked except during inspections and maintenance.

RPWT © 2008
Some hatches on elevated tanks are close to, or on, the side of the tank.
Only trained personnel with proper equipment should inspect towers.

Hatch, Manway or Bolted Panel

Access manways in the riser pipe, should area, access tube, bowl area or any other location opening directly into the water compartment should be located in strict accordance with current AWWA standards. These openings should not be less than 24 inches in diameter. However, in the case of a riser pipe or access tube of 36 inches in diameter or smaller, the access manway may be 18 inches by 24 inches with the vertical dimension not less than 24 inches. The primary access manway in the lower ring or section of a ground storage tank should be not less than 30 inches in diameter. Where necessary, for any access manway which allows direct access to the water compartment, a gasket should be used to make a positive seal when the access manway is closed.

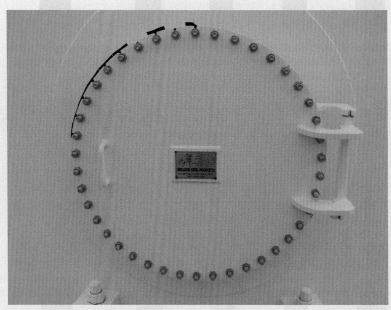
Manway hatch

Overflows should be designed in strict accordance with current AWWA standards and shall terminate with a gravity-hinged and weighted cover. The cover should fit tightly with no gap over 1/16 inch. If the overflow terminates at any point other than the ground level, it shall be located near enough and at a position accessible from a ladder or the balcony for inspection purposes. The overflow(s) should be sized to handle the maximum possible fill rate without exceeding the capacity of the overflow(s). The discharge opening of the overflow(s) should be above the surface of the ground and shall not be subject to submergence.

Overflow flapper valve

All clearwells, ground storage tanks, standpipes, and elevated tanks should be painted, disinfected, and maintained in strict accordance with current AWWA standards. No temporary coatings, wax grease coatings, or coating materials containing lead should be allowed.

No other coatings will be allowed which are not approved for use (as a contact surface with potable water) by the EPA, National Sanitation Foundation (NSF), or United States Food and Drug Administration (FDA). All newly installed coatings should conform to American National Standards Institute/National Sanitation Foundation (ANSI/NSF) Standard 61 and must be certified by an organization accredited by ANSI.

 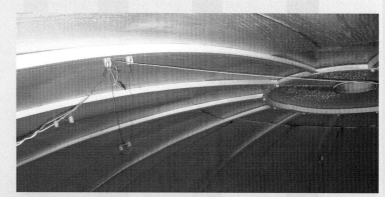

Interior walls and inside roof of a water tower

Findings of paint conditions can be graded in accordance with the applicable standards from the following agencies:

Coatings—Society for Protective Coatings ANSI/SSPC-Vis 2-82/ASTM-D6 10-85

Corrosion—National Association of Corrosion Engineers
ASM/NACE RPO 178-91 A,B,C

Welds—American Welding Society ANSI/AWS B1.11-88

Concrete—American Concrete Institute — ACI 201.1R-92

Non-destructive coating mil thickness test—DFT (dry film thickness) of installed coating system. DFT sampling should be performed on various surfaces of the reservoirs listed. Representative readings (location and findings), from accessible areas.

Is the tank in watertight condition? Potable water leaking out? Rain water leaking in?

Clearwells and potable water storage tanks should be thoroughly tight against leakage, shall be located above the groundwater table, and shall have no walls in common with any other plant units containing water in the process of treatment. All associated appurtenances including valves, pipes, and fittings shall be tight against leakage.

Water level indicator

All clearwells and water storage tanks should have a liquid level indicator located at the tank site. The indicator can be a float with a moving target, an ultrasonic level indicator, or a pressure gauge calibrated in feet of water. If an elevated tank or standpipe has a float with moving target indicator, it must also have a pressure indicator located at ground level. Pressure gauges must not be less than three inches in diameter and calibrated at not more than two-foot intervals. Remote reading gauges at the owner's treatment plant or pumping station will not eliminate the requirement for a gauge at the tank site unless the tank is located at the plant or station.

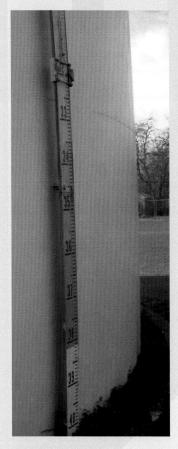

Moving target indicator   Pressure gauge

Basic Security Evaluation

Facility security; All potable water storage tanks and pressure maintenance facilities should be installed in a lockable building that is designed to prevent intruder access or enclosed by an intruder-resistant fence with lockable gates. Pedestal-type elevated storage tanks with lockable doors and without external ladders are exempt from this requirement. The gates and doors must be kept locked whenever the facility is unattended. See chapter Five

Pressure tanks

Pressure tank inspections should determine that the pressure release device and pressure gauge are working properly, the air-water ratio is being maintained at the proper level, the exterior coating systems are continuing to provide adequate protection to all metal surfaces, and the tank remains in watertight condition. **Pressure tanks provided with an inspection port should have the interior surface inspected at least once every five years.**

Photo: Pressure tank with inspection port on the end.

Photo: The interior of a pressure tank

Sanitary Surveys

A Sanitary Survey includes water storage facilities. However, it is on a much wider scope also considering the entire water system source water, cross-connections, pumps etc. The sanitary survey may include a potable water tank or tower inspection report from an inspection contractor. This will allow the inspector performing the Sanitary Survey to see the roof of the tank, vent screens, interior conditions and sediment levels without climbing the tank or tower.

In April 1999, the EPA Office of Water has issued a paper titled "Guidance Manual for Conducting Sanitary Surveys of Public Water Systems; Surface Water and Ground Water Under the Direct Influence (GWUDI)" This 182 page guide will take you through conducting a sanitary survey on your system from start to finish.

CHAPTER SEVEN

Inspection Methods

Ron Perrin with a ROV inspection camera

We have reviewed Water Personnel vs. Hiring a Inspection Contractor.

Now we should give some thought to HIGH TECH vs. LOW TECH.

Although you may think of a water utility worker with a clip board as low tech and the Inspection contractor with specialty cameras as high tech, this is not always the case.

There are many inspection contractors that will take your tank out of service and drain the water to perform the inspection. If entry is made during this process the tank must be decontaminated, this usually includes super chlorinating the water and re-filling the tank again causing the city to be out additional time and money.

Other contractors will show up with a video camera and record only the outside and above waterline conditions of the tank. Although these inspections may be very inexpensive without documenting sediment levels and the underwater conditions, there has not been much accomplished. Digital still cameras and video cameras are so common, a contractor could show up with nothing but his cell phone and accomplish this type of documentation. To be truly high tech these days you need a underwater video camera that can document the interior conditions without draining the facility or causing a disruption in service.

Photo: Ron Perrin holding a custom made underwater inspection camera and lighting system, preparing to inspect a 750,000 gallon storage tank in east Texas.

A qualified inspection contractor will have a method to inspect the facility without taking it out of service. The most common methods used are diver inspections and remote video inspections.

1.   Diver inspections are suggested when the facility is over ten million gallons. If a more detailed inspection is needed, the diver is sealed in a dry suit (make sure no one gets into your storage facility in a wet suit, more on that later). He is then washed off with a chlorine solution prior to entering the tank. The diver is then free to move about the tank to measure sediment, photograph and video interior conditions.

2.   Remote camera inspections are suggested for standpipes, and all tanks up to ten million gallons. Tanks over ten million gallons that have multiple entry points may also be inspected by using a ROV. This is safer than the diver method because a person is not entering the confined space that is, of course, even more dangerous because it is also underwater. The expert tank inspector can deliver clear sharp digital photos of the interior roof conditions and underwater video, documenting the underwater conditions and floor of the tank.

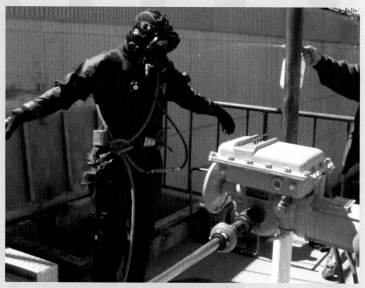

RPWT © 2007

Diver is sealed in a dry suit and washed down with a chlorine solution before entering a water tower in Southern Texas. The dry suit is much different from a wet suit that allows a layer of water between the diver and the suit. The dry suit completely seals the diver in his own environment allowing NO CONTACT WITH THE WATER AT ALL.

Whether your tank is inspected with a remote camera system, ROV, or a dive team there are a few things they must have.

A.   All Equipment entering the potable water system must be purchased for and only used in POTABLE WATER.

B.   The diver or remote camera must be washed off with a 200ppm chlorine solution to meet AWWA standards found in AWWA Standard C652-92.

C.   Insurance. The contractor should provide proof of insurance. To insure coverage have the proof by having their insurance coverage proof faxed to your office from their insurance agent naming your utility additionally insured. The liability coverage should be for at least One Million Dollars. Proof of auto and workers compensation insurance should also be provided.

D.   Proof of training. Demand to see training certification cards from a Contractors Safety Council or other reputable safety training organization.

E.   References. If they have A-D, chances are you are not the first person they have worked for. Get names and numbers of recent contracts AND CALL THEM.
References are not worth the paper they are written on if they are not checked.

RPWT © 2005
Ron Perrin Water Technologies Inspector using a remote camera to inspect a small ground tank

The Video Ray ROV is a popular inspection tool. Only 12" long, it only weighs 8 lbs.

CHAPTER EIGHT

Cleaning Methods & Procedures

A few inches of sediment being removed from a water tower by a dive crew. RPWT© 2007

Cleaning methods vary as much as inspection methods do. On the low-tech side, I have seen water utility personnel use everything from buckets and shovels to fire hose and bobcat front end loaders. These methods all require for the tank to be drained. Dewatering a water storage facility is often a nightmare.

Why is dewatering a nightmare? Once a water storage facility is drained, the loose sediment begins to dry and harden. Multiple inches of sediment that had been is a semi—liquid state and could easily be removed by a dive team, may be reduced to less than a ½ inch of hard, clay like buildup that must be scraped from the floor of the tank. The painted surface of the tank floor is much better off if the sediment is removed before it is allowed to dry and get hard.

Sediment being removed by dive team RPWT © 2008

Diver entering tank with SCUBA (Self-Contained Underwater Breathing Apparatus) RPWT© 2008

The blue bottle on his back is his primary air supply, the smaller bottle on his side is his back up air. It is positioned so he can turn the valve on if needed. This insures that the emergency air will not leak through the regulator during the dive. This is basically the same type of back up system cave divers use.

The diver is using line-surfaced supplied air with 30 cubic feet of compressed air in a bail out bottle for emergency back up.

The mask being used on this dive is a Kirby Morgan EXO. This is a full face mask that is also equipped with diver to surface communication. He is also tethered to meet OSHA 29CFR requirements All equipment used inside the potable water storage tank is purchased for, and only used in, potable water storage facilities. Before entering the tank the diver is washed down with a 200ppm chlorine solution to meet AWWA & TCEQ requirements.

The best method for cleaning your tank is to use a qualified and experienced dive team.

a.   Removes all loose sediment

b.   Will not damage tank

c.   Little if any down time, because the water is not drained the tank goes back in service quick

d.   Small tanks can be cleaned in hours instead of days. Large tanks can be cleaned in days instead of weeks

Once underwater, the diver carefully works off the ladder to clean a spot to stand.

He then works slowly outward from that spot, careful not to stir up the sediment and create turbidity. This is one of the big differences between a professional tank cleaning crew and a commercial dive crew that has been trained to work offshore. They tend to throw everything they need into the tank, and once the sediment is disturbed it becomes a big cloud in the water. This turbidity has a negative effect on the water system. The dirty/cloudy water can then get into the system causing complaints. In addition to that, the tank can never really be cleaned on that day because the sediment that is suspended in the water causing the cloudy and high turbidity condition will eventually settle down to be another layer of sediment on the floor of the tank. There is an art to cleaning a potable water storage tank without creating turbidity.

A few inches of sediment being removed from a water tower by a dive crew. RPWT© 2007

CHAPTER NINE

Hiring an Inspection or Cleaning Contractor

The importance of hiring contractors who obey the rules-

OSHA rules and regulations may seem overwhelming at times, but for the most part they have been put in place due to death or injury on job sites. Each year, workers die on construction sites. Inspecting water storage tanks is a dangerous job. Many utilities do not want their employees climbing tanks and towers due to the risk of falling, so this is often why contractors are hired to inspect them. Other risks are also present and should not be overlooked. Two high ones are due to the storage area being a confined space and the fact that chemicals are used to heat the water, so that the quality of air and oxygen content becomes an issue. This is a primary reason why I will use a remote camera or ROV if those methods will provide enough information to make my customer happy.

Lockout Tag out Procedures-

Locking out pumps when inspecting clearwells and ground tanks at water plants is extremely important. Failure to do so may have deadly consequences as it did in on May 12, 2008. A diver was inspecting a tank at the Paris, Texas water plant. According to the Paris News in an article published May 20, 2008:

> " . . . he drowned when the storage tank pumps turned on, pinned him and sucked his breathing apparatus away . . . . The pumps were off when he went into the tank," said Paris City Manager Kevin Carruth. "But the pumps are automatic and turned on while he was diving." The accidental drowning still is under investigation by local authorities and by OSHA. Meanwhile, his Coast Guard friends are closely watching the investigation."

Region 6 News Release: OSHA 08-1656-DAL
Nov. 10, 2008

OSHA cited the company with two alleged willful violations and three alleged serious violations. The willful violations involve failure to conduct a safety and health assessment for surface and underwater conditions when planning diving operations and to brief dive team members of any hazards or environmental conditions that may affect the safety of the diving operation. The serious violations involve entry into a confined space without verification it was safe for entry; the employer's failure to ensure that the supervisor verified all tests were conducted, and all equipment and procedures were in place in accordance with confined space requirements; and failure to develop and document procedures for the control of potential hazardous energy.

The citations carry $64,400 in proposed penalties.

A willful citation is issued for violations committed with disregard of or plain indifference to the requirements of the Occupational Safety and Health Act and regulations. OSHA defines a serious violation as a condition that exists where there is a substantial possibility (of) death or serious physical harm can result.

U.S. Department of Labor releases are accessible on the Internet at www.dol.gov.

When our company was hired to finish the job left undone, we were given a few more details. The diving company had been the low bidder in 2008. They had offered a 3 man dive crew for less than our 1 man remote camera inspection. On paper this looks to be a much better deal, yet safety should always come before speed. We were told the dive crew entered the water plant after hours without letting anyone know they were there. The operator in the plant had no idea there was a dive crew working in a tank. OSHA did not cite the city or plant operator on duty because of this reason.

OSHA guidelines regarding **Lockout/Tagout**:

STD 01-05-019—STD 1-7.3—29 CFR 1910.147, the Control of Hazardous Energy (Lockout/Tagout)—Inspection Procedures and Interpretive Guidance

Lockout/Tagout (LOTO) is addressed in specific standards for the general industry, marine terminals, longshoring, and the construction industry.

Read more here: http://www.osha.gov/SLTC/controlhazardousenergy/standards.html

OSHA
Section 5(a)(1) of the OSH Act, often referred to as the General Duty Clause, requires employers to "furnish to each of his employees employment and a place of employment which are free from recognized hazards that are causing or are likely to cause death or serious physical harm to his employees". Section 5(a)(2) requires employers to "comply with occupational safety and health standards promulgated under this Act".

Also see:

http://www.osha.gov/pls/oshaweb/owadisp.show_document?p_id=9804&p_table=STANDARDS

1910.147(a)
Scope, application and purpose—
1910.147(a)(1)
Scope
1910.147(a)(1)(i)

This standard covers the servicing and maintenance of machines and equipment in which the **unexpected** energization or start up of the machines or equipment, or release of stored energy could cause injury to employees. This standard establishes minimum performance requirements for the control of such hazardous energy.

Ron Perrin Water Technologies Field Technician ascending a water storage tower in North Texas. RPWT© 2008

Low vs. High Tech Inspection Needs

There are several differences between low and high tech inspections, and choosing the right one can make a difference in a job well done.

The low tech inspection may include a man with a 12 point check list on a clip board. This may meet requirements, but to check sediment levels the tank is often completely or partially drained. Entering a confined space without a breathing apparatus that possibly has contaminants listed in chapter four is not high on my list of things to do. Without a fresh air supply there is too much risk in this type of a confined space. This is one reason I prefer to send a camera in to see what is going on. My second choice, if more information is needed or if the tank needs to be cleaned out, is a potable water diver. The diver has his primary and secondary air supply, so he is never at risk for breathing in contaminants that may be in the confined space of a water storage tank or tower.

If your tank is less than 20,000 Gallons, you may be able to see the floor and estimate the sediment levels without draining the tank or any additional assistance.

For larger facilities, the low tech method has numerous disadvantages. Using an underwater video camera to document the inside roof wall areas and floor would be considered a high tech inspection method. This kind of inspection gives you a much better idea of what is going on inside the tank. It also allows you to have hard documentation in the form of photos & video of what condition your facility was in at this point in time.

We still see some engineering firms and a lot of painting contractors remove tanks from service and have the water utility personnel drain the tanks. This seems backwards with the current technology available today.

I would also put a person with a common video camera in a low-tech category. Beware of proposals that claim to quote a "VIDEO" Inspection. If they do not say an UNDERWATER VIDEO, or, even better these days, an UNDERWATER DVD, you may end up having to pay for an inspection that fails to show you anything under the water line or having to take your tank out of service so it can be inspected.

Another common inspection feature that you will find most professional tank inspection companies offer is DFT testing. DFT stands for **D**ry **F**ilm **T**hickness testing. This can be done on the outside of the tank where the paint is dry. It provides a measurement to see how thick the paint actually is. Ask your prospective tank inspector if they perform DFT testing. If they do not know what DFT stands for this may be a clue.

Unless you have a five million gallon tank or even a 2 or 3 million gallon tank with only one opening, I prefer the remote camera. You are able to get a lot of information with a remote camera system in the hands of a trained inspector. The remote camera is an underwater camera that is simply lowered into a storage tank to record the interior roof, walls, and floor. For larger ground tanks and underground clearwells, a ROV allows you to maneuver to the far ends of the facility for a better inspection.

Photo: Ron Perrin & his son Bradley preparing to inspect
An underground water storage tank with his Video Ray ROV system.

Points to remember when hiring any Potable Water Tank Inspector—

A. Anyone climbing on a tank or tower should have proper safety equipment to be tied off at all times. All climbing equipment should be in good condition and have an inspection record attached to it.

B. All inspection equipment entering the potable water system must be purchased for and only used in POTABLE WATER.

C. If a remote camera or any other type of equipment enters your potable water it must be washed off with a 200ppm chlorine solution to meet AWWA guidelines. Any contractor should know this (great test question).

D. Insurance—Have the contractor provide proof of insurance by having their insurance coverage faxed to your office from their insurance agent naming your utility additionally insured. The liability coverage should be for at least One Million Dollars. Proof of auto and workers compensation insurance should also be provided.

E.   Proof of training. Demand to see training certification cards from a Contractors Safety Council or other reputable safety training organization.

F.   References. If they have A-D, chances are you are not the first person they have worked for. Get names and numbers of recent contracts AND CALL THEM.
References are not worth the paper they are written on if they are not checked. Confirm that the former customers were able to see sediment levels, NOT JUST THE SURFACE OF THE WATER. You need to know you will see the floor of the tank, the side wall condition, and the inside roof.

Diving Contractors

When considering different bids from diving contractors, look for a contractor who is in business to serve the potable water industry.

Look at the company web site and literature. If it is obvious that a contractor spends most of his time working off shore for the oil and gas industry they may not be a good choice to enter your potable water tank. Remember all equipment must be purchased for and only used in POTABLE WATER. If they use their gear off shore or in lakes and rivers it is prohibited from being used in potable water and may cause some serious health concerns.

Do not assume he knows what he is doing just because he has a SCUBA certification or even a Commercial Diving Certification from the Association of Diving Contractors (ADC). Diving in potable water storage tanks presents unique challenges to the traditional diving industry.

Do they may meet OSHA requirements for training and experience under TITLE 29—LABOR PART 1910 Sec. 1910.410, Qualifications of dive team? It states "Each dive team member shall have the experience or training necessary to perform assigned tasks in a safe and healthful manner." Most would agree that OSHA has set the bar a bit low here. There are a few things you want to ask for to confirm they know the rules of TITLE 29—LABOR PART 1910, AWWA, EPA & State guidelines.

Here are some important questions to ask when selecting a diving contractor.

A.   Ask where his most recent jobs have been. All Equipment entering the potable water system must be purchased for and only used in POTABLE WATER.
I have a detailed list about dive gear listed below in "Things to check when the dive crew arrives on site, make a check list from the items."

B.   Ask what, if any, decontamination steps they take prior to putting a diver into your water system. If they wash down with chlorine, ask what the concentration of their chlorine solution is. Remember the diver and all equipment must be washed off with a 200ppm chlorine solution to meet AWWA and most state guidelines.

C.   Ask what will separate the Diver from the Water?

They should have a Dry Suit, NOT A WET SUIT.

We have actually caught competitors using wet suits in potable water storage systems. Here is the difference: the dry suit keeps the diver separated from the water. He actually dives in his street clothes, takes the dry suit off at the end of the day, and goes home. A wet suit allows water to get between the diver and the suit. The divers' body warms up the water that slowly circulates in and out of the suit while he is underwater. Wet suits are designed for sport diving and only cost about 10% of what a commercial dry suit would cost. After the dive he is soaking wet, usually in a swim suit. If you see someone about to get into your tank and they change into a swim suit before putting on their dive gear you should ask if he is planning to get wet while in your tank. There is NO reason for a professional potable water diver to EVER come in contact with the water. See more on this in the next section.

D. Insurance—Have the contractor provide insurance proof by having their insurance coverage proof faxed to your office from their insurance agent naming your utility additionally insured. The liability coverage should be for at least One Million Dollars. Proof of auto and workers compensation insurance should also be provided.

E. Proof of training. Demand to see training certification cards from a Contractors Safety Council or other reputable safety training organization. Diving certifications are issued from a wide range of agencies. I would question basic or "Open Water" certifications. However with the proper experience these would meet OSHA 29 CFR requirements. You may review OSHA 29CFR in the rear of this book.

F. Get a typed Reference list with dates and phone numbers and compare it to his verbal reply to question A. References are not worth the paper they are written on if they are not checked, so CHECK THEM.

G. Safety Issues;

Ask if they have a Safe practices manual that includes:

1 LOCK out And TAG out procedures
"The control of hazardous energy (lockout/tagout).—1910.147"
Locking out pumps when cleaning or inspecting a water plant should go without saying. This should be listed in their "Safe practices manual". (As recently as May of 2008, a diver was killed inspecting a water plant in Northeast Texas.)

2 Confined space procedures.
(Confined Spaces [29 CFR 1910], page 282 dated August, 1985), A confined space permit should be issued by the contractor prior to a diver entering into a potable water storage tank. Procedures should be listed in their "Safe practices manual".

Things to check when the dive crew arrives on site, make a check list from the items:

1.   Ask to see their "Safe practices manual" they should have on site as required in TITLE 29—LABOR PART 1910.

2.   Ask to see the Confined Space entry permit. **CFR 1910.146 (f)(11)** See appendix to read the entire regulation and to see an example from GSA.

3.   Ask to see diver certifications , first aid & CPR CARDS.

<u>1910.410(a)(3)</u>

All dive team members shall be trained in cardiopulmonary resuscitation (CPR) and first aid (American Red Cross standard course or equivalent).

4.   Ask to see their DRY SUIT-

As previously stated: The difference between a dry suit and a wet suit is crucial when diving in potable water. A "WET SUIT" allows a layer of water to get between the diver and the suit. The person warms up this layer of water allowing him to be warmer than he would be without a wet suit. Some wet suits are very thick allowing more warmth but all wet suits allow water to flow in and out over the divers' body. A bathing suit is typically worn under the wet suit because the person using this equipment knows he is going to get 100% wet. From a health standpoint, an industrial hygienist would probably consider a person diving in a "wet suit" about the same as a naked man in your water supply.

The **"DRY SUIT"** keeps the diver dry. Typically he wears his street clothes under the dry suit because he or she is not planning to get wet.

5.   Ask to see the dive mask or helmet –

A full face mask, band mask or diving helmet should also be used. Most people know what a normal skin-diving mask looks like; it covers the eyes and nose. The full face mask, band mask or diving helmet covers the entire face. Along with a hood on the Dry suit, there should be no part of the divers body exposed at all.

6.   Ask about their chlorine wash. You want to know how the diver is going to be washed down before entering the tank and what the strength of the wash is.

Once the diver is sealed into his own environment, the diver and all equipment entering the water storage tank should be washed down with a 200ppm chlorine solution to meet AWWA standards. All equipment used in a potable water tank should be purchased for and only used in potable water

7.   Fall Protection—**1926.500(a)(2)(vii)** Requirements relating to fall protection for employees working on stairways and ladders are provided in subpart X of this part.

Contractors working on Water Tanks & Towers Should have:

Personal fall arrest system—A system used to arrest an employee in a fall from a working level. It consists of an anchorage, connectors, a body belt or body harness, and may include a lanyard, deceleration device, lifeline, or suitable combinations of these. As of January 1, 1998, the use of a body belt for fall arrest is prohibited. The harness should have an inspection record attached to it that shows the equipment has been regularly inspected.

Deceleration device—A mechanism, such as a rope grab, rip-stitch lanyard, specially-woven lanyard, tearing or deforming lanyards, automatic self-retracting lifelines/lanyards, etc., which serves to dissipate a substantial amount of energy during a fall arrest, or otherwise limit the energy imposed on an employee during fall arrest.

Lanyard—A flexible line of rope, wire rope, or strap which generally has a connector at each end for connecting the body belt or body harness to a deceleration device, lifeline, or anchorage.

1926.500(a)(1) This subpart sets forth requirements and criteria for fall protection in construction workplaces covered under 29 CFR part 1926. Exception: The provisions of this subpart do not apply when employees are making an inspection, investigation, or assessment of workplace conditions prior to the actual start of construction work or after all construction work has been completed.

1926.500(a)(2)(v) Requirements relating to fall protection for employees engaged in the erection of tanks and communication and broadcast towers are provided in **§ 1926.105**.

8.   **Hard Hats** – For years we had operated with the ground person using a hard hat and the climbers using just caps or nothing at all for the most part. Hard hats worn by the climber usually ended up falling, creating a bigger safety hazard.

Be sure the contractor has a hard hat for at least every person on the ground crew.

You can also read in the regulation below it says "each affected employee" wears a protective helmet not hat. The people on the ground under a water tower are affected. Persons without hard hats need to be told to leave the area. For additional safety the climber should wear a helmet.

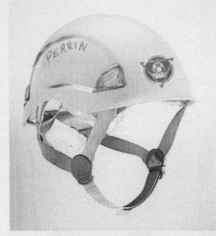

Helmets like this one have a Chin Strap. It decreases the risk of losing the helmet in the event of a fall. More importantly to me, since I am usually on the ground, it stays on the climbers' head and won't blow or fall onto me.

1910.135(a)(1)

The employer shall ensure that each affected employee wears a protective helmet when working in areas where there is a potential for injury to the head from falling objects.

1910.135(a)(2)

The employer shall ensure that a protective helmet designed to reduce electrical shock hazard is worn by each such affected employee when near exposed electrical conductors which could contact the head.

1910.135(b)

Criteria for protective helmets.

Diver prepares to enter a water tower RPWT© 2007

Diver prepare to enter tank at water plant to remove
a costic build up in a 20" line RPWT© 2007

Costic build up removed from a 20" line **RPWT © 2007**

A good dive crew specializing in potable water diving can perform a wide variety of services for a water utility. In addition to inspections and cleanings, they can be used to set plugs on the side of ground tanks to allow valves to be changed with no water loss and also to remove costic build up in water lines as seen in the photos above.

Checking Safety Records

Checking safety records only takes a few minutes, if you know where to look. The OSHA site : www.dol.gov/compliance will allow you to search for violations in a number of ways. The best way I have found is to put OSHA NEWS RELEASE and the name of the company you want to check on into Google.

Ethics-

Ethics could be an entire chapter or even its own book. I had to say something about ethics in this chapter because of what I have witnessed in the potable water diving industry.

Since 1992 I have been involved in the inspection and cleaning of potable water systems.

However, my formal training was in criminal justice. I was originally commissioned as a Texas peace officer in 1984 and worked at it full time through 1993. I continued to serve as a volunteer officer, and by 2003 I had advanced to a certified police instructor and held a Master Peace Officer certification. I logged well over 1500 training hours during my law enforcement career. The two courses of which I am most proud of were Field Instructor for Louisiana State University Academy of Counter Terrorist Education Instructor which allowed me to instruct officers on how to deal with terrorist attacks. The other course was at the Institute for Law Enforcement Administration on May 9th-13th 2006 to be certified to teach ethics to police officers.

But you don't need to be an ethics instructor to know the difference between what is right and what is wrong.

While attending the Texas Water Utilities Association annual conference at A&M University several years ago, a water utility superintendent from the DFW metroplex approached me with a small problem.

He stated; "Last year I had a diving company come out and inspect my water tanks. We had a lot of sediment in the towers but most notably there was a Beer Can. So I hired that company to clean the tanks. After the cleaning was done they gave me a video of a clean floor. This year I had them back again to do the inspections. Guess what? They showed me a video with that same beer can. Now, how can that be?"

The answer to him was: "It can't be."

As long as I have been inspecting water storage tanks & towers this is the only beer can I have heard about inside a tank. To say this is a rare find would be an understatement. The company had to do one of two things:

The tank was not cleaned and a video of a different clean water tower floor was delivered with the cleaning video.

The tank may not have been inspected the second time. They could have used the old inspection tape again. This may have been done because the tank was not properly inspected the second time, they lost the video, or even worse, they wanted to show a dirty tank to get another cleaning job.

With this known, the city kept hiring this un-ethical contractor for years.

In another case, I had a diving contractor I worked with for years who specialized in potable water cleaning. He told me about working for a company that told him to clean

½ of a water storage tank and video tape it twice so it would look like the entire floor had been cleaned. In another case, they (the owner of the company) told him to clean the areas under the vent structures and under the entry hatch area so if anyone tried to look into the tank they would not be able to see the areas that had not been cleaned.

The employee had more ethics than the company so he resigned and started his own diving service. He went back to the utilities that he knew were cheated and told them. The response was a shock to him. The water superintendents decided to take no action because they would have looked foolish for allowing this to have happened and feared losing their jobs.

Another story that particular contractor related to me has stuck with me for years. It was about a water utility superintendent working for a city in Northeast Texas. After years of repeated requests to get funding to have his tanks inspected he was finally granted the purchase order.

The inspection showed the tanks were in such poor condition you could see daylight coming through the metal sidewall near the high water mark inside the tank. The tank was so corroded that the diving contractor feared that the suction from his cleaning pumps may have sucked a hole through the metal. The superintendent took the report to the City Council. He was terminated for allowing the water storage tank to get in that condition.

Once, I was asked to present my inspection report at a city council meeting in a small town in central Texas. The director of public works had asked for funding to clean the cities water storage tanks and had been turned down repeatedly. He was hoping that by having me present my report as an independent consultant would educate the council on how desperately the water storage tanks actually needed to be cleaned. The council refused to watch my video. The next time I visited that community the director of public works had been demoted to the dog catcher. Ethical?

These days our diving contracts are completed by my own employees giving me more control over the quality of service. I hired a diver that worked for a well known established company from northern part of the Untied States. He was well trained with impressive credentials and training and eager to show us how efficient he was. The speed at which he could clean a water storage tank came at a high cost. He didn't last long with us because he would not take the time to remove all of the sediment from the floor of the tank. He would claim to be finished, but I would have another person check his work. We would find that we would have to go back and re-dive to actually remove all of the sediment from the floor of the facilities. He had been trained by the other company to clean 2 water towers a day. This is possible to do if the sediment is light enough. If there is a lot of sediment however, it may not be possible to finish a tower cleaning in one day.

The other company gave a bonus at the end of the month if a certain dollar amount was reached by each cleaning crew. To get the bonus, you only had a limited amount of time to spend on any one tank. So the company says clean the tanks—stay till it's done, but then sets up a bonus payment that limits the amount of time you can spend on any one tank. In this case, the ethics of the company policy affects the ethics of the divers by rewarding them for moving faster; rewarding them for putting speed before ethics.

Here is the thing about ethics. For a water utility or purchasing department to make an ethical selection, it will more often than not be a more expensive selection. When you're talking about peoples' lives, that is money well spent. Yet, anyone can say that they are ethical. Honesty and ethics **is not free** when living by ethical principals.

You will not always be better off for doing the right thing. I'm sorry if I am the first one to tell you, but life isn't always fair, and doing the right thing may not always be to your benefit. It's done because it's right. Have you heard the term "time is money"? When unexpected things happen on a jobsite, it costs time. That time equals money, and it sometimes cost my profit margin on the job, and on some jobs I don't get close to breaking even. Giving the crew the time to do the job completely, and taking the time to be safe, is the biggest challenge for any contractor. When you are dealing with high risk contracting on top of the financial aspect, you are dealing with peoples' lives when you don't give them the time to be safe.

Dealing with ethics in today's world is a problem in every industry.

Values in America are changing our moral fabric, and the ethics once lived by are suffering. Consider these statistics from Patterson and Kim, "The Day America Told the Truth" 1991.

A survey of Americans found the following:

—Believe in All 10 commandments   13%
—Call in sick when they are not   50%
—Lie regularly (work and home)   91%
—Don't know their next door neighbors   70%

Ethics, or lack of ethics, has become a cancer in our society. Fraud, embezzlement, and misuse of public money are common stories on the nightly news. It affects every aspect of our lives.

Ethics comes down to what we do. Whether we are a judge, police officer, water utility worker, or a contractor working for a water utility, those of us who make our living protecting the public should be held to a higher standard. When hiring a contractor to perform work for a public utility, the lowest bid should not be the only standard.

ETHICS & the LOW BID CONTRACTOR

I call these guys the **lowballers.** Let me explain who I am calling a lowballer. It's not a company that bids a few hundred dollars less than the next guy; it is the company that put a bid of $5,000 when all others are over $8,000. An ethical contractor has no chance against the lowballers. They are going to get the contract for a third of the going rate. In any other industry, a bid that is too low is dismissed. Municipal governments that are so damaged by "good old boy" politics and "brother-in-law" deals have a set policy, or even ordinances, that state the LOW BID MUST PREVAIL. This kind of thinking must be re-thought for some contracts.

Unlike the purchase of materials, when dealing with HIGH RISK CONTRACTORS, factors other than price must be considered. Automatically taking the low bid when dealing with any high risk contractor is not a good policy. How much time is the project going to take? Is their low bid enough even to cover the minimal time to do the job? You should do more than look at the dollars bid. Call each contractor and ask them to quote a daily rate for a dive team. Then ask them how long your job will take. You may learn that the lowest bidder has not allowed enough time to complete the job. Can they really do the job, pay their employees, cover fuel, hotel, and other expenses and still make a modest profit for the company on 1/3, or even ½, of what others bid?

THE TRICK of the FOLDING YARD STICK & the VIDEO DOUBLE BACK.

Over the years I have hired a lot of divers, and I have heard stories of how companies push ethics aside to go after the biggest check possible in the shortest amount of time. One company used a folding yard stick for years. They would break off the last part of the stick so when they took a video of the ruler sticking into the sediment it was always at least 9 inches. Another company was in a habit of cleaning half a tank and when it came time to perform the after video they just doubled back and filmed the clean part twice. Some companies tell their employees to clean the entire tank, and then offer bonuses if multiple tanks are cleaned in a single day. This sets up the divers up to be rewarded if they move fast, leaving before the tanks are really done, and go on to clean, or at least appear to clean, more tanks. Rewarding divers for speed is a doubled edged sword. On one side the crew definitely moves faster, but on the other, the utility is often left with a job half done.

Unfortunately, old habits are hard to break. When these divers tried to take shortcuts like this with our company, they were shocked to learn that the after cleaning video was done by another person, and when sediment was still found in their tanks that were reported to be clean, there were dire consequences. This problem is so wide spread that despite experienced workers having the expertise to properly do the job we have had much better results when we hire raw recruits and train them from scratch. I spoke about money well spent when you're talking about peoples' lives, which is not only about the divers having the time they need to be safe. The contracts I am referring to in this chapter are made to remove sediment from the floor of water storage tanks. Potable water goes directly to the public. If only part of the tank is cleaned, bacteria and other contaminants may continue to grow in the tank and threaten the health or even the lives of the customers on that water system (see Chapter four). Spending the money it takes to get the job done right is money well spent.

An extremely low bid should raise suspicion. Check references, check safety records, investigate to find out what their daily rate is, and how much time they are really planning to be on your site. Do what many general contractors do, get more bids. Throw out the high bid, throw out the low bid, check safety records and references of the ones in the middle, and make a selection.

ETHICS & SAFETY:

Take a little time to make to sure the contractors you hire have good safety records. Search engines like Google have made checking a company's safety record easier than ever before.

Safety and good ethics go hand in hand. Providing employees the time they need to perform their jobs safety and providing the proper equipment and training all cost money. Safety records and references are things that you can check to get an idea of the policies and customs of the company. Repeated or willful violations on safety issues may reflect a company that is more concerned about making a buck than doing what is right.

A company's safety record must be considered when hiring a HIGH RISK CONTRACTOR.

There is more than one way to award a contract. There are two commonly used methods to determine who is awarded a contract. The low-bid method and the qualifications-based process.

The low-bid method that is by far the most common method of selecting an inspection or diving contractor. Qualifications-based process is usually reserved for attorneys, engineers, and architects. When is the last time you asked your doctor for a bid on doing a little surgery?

Architects and engineers for the most part avoid getting into a low bid competition. Professionals services are more likely to use a qualifications-based process that allows the buyer to choose a contractor based on the contractor's qualifications, experience, and perhaps most importantly, reputation.

Perhaps it is more ethical to choose the qualifications-based process when people are hired to climb water towers, crawl into confined spaces, and work underwater.

These jobs require highly trained, well equipped people to tame the high risks that are associated with climbing towers and working underneath 40 feet of water. These contractors are specialized and much closer to professional status than their blue collar counterparts who are awarded skilled trade or typical construction type contracts. If you need a street paved, a roof installed, or a ditch dug, there is merit in seeking out the low bid contract. However, these can come with their own risks as well.

Tower inspectors and commercial divers have more at risk when they go to work, and the water utility has an ethical obligation to use the qualifications-based process when dealing with this type of contract. With this method, the utility can choose a contractor based on the contractor's qualifications, experience, reputation, and perhaps most importantly, with these types of contracts it allows you to evaluate the company's safety records.

INDEX

Helpful Definitions:

What is a Public water system?

A public water system—A system for the provision to the public of water for human consumption through pipes or other constructed conveyances, which includes all uses described under the definition for drinking water. Such a system must have at least 15 service connections or serve at least 25 individuals at least 60 days out of the year. This term includes; any collection, treatment, storage, and distribution facilities under the control of the operator of such system and used primarily in connection with such system, and any collection or pretreatment storage facilities not under such control which are used primarily in connection with such system. Two or more systems with each having a potential to serve less than 15 connections or less than 25 individuals but owned by the same person, firm, or corporation and located on adjacent land will be considered a public water system when the total potential service connections in the combined systems are 15 or greater or if the total number of individuals served by the combined systems total 25 or greater at least 60 days out of the year. Without excluding other meanings of the terms "individual" or "served," an individual shall be deemed to be served by a water system if he lives in, uses as his place of employment, or works in a place to which drinking water is supplied from the system.

Community water system—A public water system which has a potential to serve at least 15 residential service connections on a year-round basis or serves at least 25 residents on a year-round basis.

AWWA standards—The latest edition of the applicable standards as approved and published by the American Water Works Association, 6666 West Quincy Avenue, Denver, Colorado 80235.

Contamination—The presence of any foreign substance (organic, inorganic, radiological or biological) in water which tends to degrade its quality so as to constitute a health hazard or impair the usefulness of the water.

Disinfectant—Any oxidant, including but not limited to chlorine, chlorine dioxide, chloramines, and ozone added to the water in any part of the treatment or distribution process, that is intended to kill or inactivate pathogenic microorganisms.

Disinfection—A process which inactivates pathogenic organisms in the water by chemical oxidants or equivalent agents.

Drinking water—All water distributed by any agency or individual, public or private, for the purpose of human consumption or which may be used in the preparation of foods or beverages or for the cleaning of any utensil or article used in the course of preparation or consumption of food or beverages for human beings. The term "Drinking Water" shall also include all water supplied for human consumption or used by any institution catering to the public.

Drinking water standards—The commission rules covering drinking water standards in Subchapter F of this chapter (relating to Drinking Water Standards Governing Drinking Water Quality and Reporting Requirements for Public Water Systems).

Elevated storage capacity—That portion of water which can be stored at least 80 feet above the highest service connection in the pressure plane served by the storage tank.

Emergency power—Either mechanical power or electric generators which can enable the system to provide water under pressure to the distribution system in the event of a local power failure. With the approval of the executive director, dual primary electric service may be considered as emergency power in areas which are not subject to large scale power outages due to natural disasters.

Groundwater—Any water that is located beneath the surface of the ground and is not under the direct influence of surface water.

Human consumption—Uses by humans in which water can be ingested into or absorbed by the human body. Examples of these uses include, but are not limited to drinking, cooking, brushing teeth, bathing, washing hands, washing dishes, and preparing foods.

Intruder-resistant fence—A fence six feet or greater in height, constructed of wood, concrete, masonry, or metal with three strands of barbed wire extending outward from the top of the fence at a 45 degree angle with the smooth side of the fence on the outside wall. In lieu of the barbed wire, the fence must be eight feet in height. The fence must be in good repair and close enough to surface grade to prevent intruder passage.

National Fire Protection Association (NFPA) standards—The standards of the NFPA 1 Batterymarch Park, Quincy, Massachusetts, 02269-9101.

Sanitary survey—An onsite review of the water source, facilities, equipment, operation and maintenance of a public water system, for the purpose of evaluating the adequacy for producing and distributing safe drinking water.

Sediment—Sediment accumulation occurs within storage facilities due to quiescent conditions which promote particle settling. Potential Water quality problems associated with sediment accumulation include increased disinfectant demand, microbial growth, disinfection by-product formation, and increased turbidity within the bulk water.

Spalling—Concrete spalling is a common occurance, especially on exterior surfaces that are exposed to freeze and thaw cycles. It's like scaling, but involves bigger chunks breaking loose for no apparant reason. There are many causes such as improper finishing methods and curing methods. Simple repairs can be done fill the void and prevent further deterioration. oxidize or—dise Verb [-dizing,—dized] or—dising,—dised to react chemically with oxygen, as in burning or rusting oxidization—disation n
Are aircraft warning lights required on your water tower?

Code of Federal Regulations Title 14 Aeronautics and Space

§77.23 Standards for determining obstructions.

(a) An existing object, including a mobile object, is, and a future object would be, an obstruction to air navigation if it is of greater height than any of the following heights or surfaces:

(1) A height of 500 feet above ground level at the site of the object.

(2) A height that is 200 feet above ground level or above the established airport elevation, whichever is higher, within 3 nautical miles of the established reference point of an airport, excluding heliports, with its longest runway more than 3,200 feet in actual length, and that height increases in the proportion of 100 feet for each additional nautical mile of distance from the airport up to a maximum of 500 feet.

(3) A height within a terminal obstacle clearance area, including an initial approach segment, a departure area, and a circling approach area, which would result in the vertical distance between any point on the object and an established minimum instrument flight altitude within that area or segment to be less than the required obstacle clearance.

(4) A height within an en route obstacle clearance area, including turn and termination areas, of a Federal airway or approved off-airway route, that would increase the minimum obstacle clearance altitude.

(5) The surface of a takeoff and landing area of an airport or any imaginary surface established under §77.25, §77.28, or §77.29. However, no part of the take-off or landing area itself will be considered an obstruction.

(b) Except for traverse ways on or near an airport with an operative ground traffic control service, furnished by an air traffic control tower or by the airport management and coordinated with the air traffic control service, the standards of paragraph (a) of this section apply to traverse ways used or to be used for the passage of mobile objects only after the heights of these traverse ways are increased by:

(1) Seventeen feet for an Interstate Highway that is part of the National System of Military and Interstate Highways where overcrossings are designed for a minimum of 17 feet vertical distance.
(2) Fifteen feet for any other public roadway.
(3) Ten feet or the height of the highest mobile object that would normally traverse the road, whichever is greater, for a private road.
(4) Twenty-three feet for a railroad, and,
(5) For a waterway or any other traverse way not previously mentioned, an amount equal to the height of the highest mobile object that would normally traverse it. [Doc. No. 10183, 36 FR 5970, Apr. 1, 1971]

Cryptosporidium—is a protozoan parasite found in humans, other mammals, birds, fish, and reptiles. It is common in the environment and widely found in surface water supplies (Rose, 1998; LeChevallier and Norton, 1995; Atherholt et al., 1998; EPA, 2000a). In the infected animal, the parasite multiplies in the gastrointestinal tract. The animal then excretes oocysts of the parasite in its feces. These oocysts are tiny spore-like organisms 4 to 6 microns in diameter (too small to be seen without a microscope), which carry within them the infective sporozoites. The oocysts of Cryptosporidium are very resistant to adverse factors in the environment and can survive dormant for months in cool, dark conditions such as moist soil, or for up to a year in clean water. When ingested by another animal they can transmit the cryptosporidiosis disease and start a new cycle of infection.

Cryptosporidiosis is primarily a waterborne disease, but has also been transmitted by consumption of contaminated food, unhygienic diaper changing practices (and other person-to-person contact), and contact with young farm animals.

Cryptosporidium oocysts are not easily killed by commonly-used disinfectants. They are relatively unaffected by chlorine and chloramines in the concentrations that are used for drinking water treatment. Oocyst infectivity appears to persist under normal temperatures, although oocysts may lose infectivity if sufficiently cooled or heated (USEPA, 2000a).

Research indicates that oocysts may remain viable even after freezing (Fayer and Nerad, 1996).

Physical removal is critical to the control of Cryptosporidium because it is highly resistant to standard disinfection practices. Cryptosporidiosis, the infection caused by Cryptosporidium, may manifest itself as a severe infection that can last several weeks and may cause the death of individuals with compromised immune systems.

Cleaning the loose sediment from your water storage tanks removes the habitat where bacteria, protozoa, and viruses can hide from treatment chemicals, thrive and grow. Of these **Cryptosporidium** may be the worst.

Reference:

The Safe Drinking Water Act http://www.epa.gov/safewater/The SDWA/index.html

Finished Water Storage Facilities http://www.epa.gov/safewater/disinfection/tcr/pdfs/whitepaper_tcr_storage.pdf

Centers for Disease Control and Prevention http://www.cdc.gov/crypto/

National Primary Drinking Water Regulations: Long Term 1 Enhanced Surface Water Treatment Rule http://www.epa.gov/EPA-GENERAL/2002/January/Day-14/g409.htm

The Paris News "Diver's life cut short in accident" Published May 20, 2008.

Confined Space Alert—DHHS (NIOSH) Publication No. 86-110

National Assessment of Tap Water Quality by **Environmental Working Group** (ENG). *EWG is a nonprofit research organization based in Washington, D.C., that uses the power of information to protect human health and the environment. The group's work on water quality is available at www.ewg.org/issues/siteindex/issues.php?issueid=5006*

References from the *ENG National Assessment of Tap Water Quality* :
EPA (Environmental Protection Agency). 1998. Chemical Hazard Data Availability Study. What Do We Really Know About the Safety of High Production Volume Chemicals? http://www.epa.gov/chemrtk/hazchem.pdf.

EPA (Environmental Protection Agency). 2000. 2000 National Water Quality Inventory. National Water Quality Report to Congress under Clean Water Act Section 305(b). http://www.epa.gov/305b/2000report/. Accessed December 13, 2005.

EPA (Environmental Protection Agency). 2001a. Controlling Disinfection Byproducts and Microbial Contaminants in Drinking Water. Chapter 2: A review of drinking water regulations in the U.S. EPA 600R01110. http://www.epa.gov/nrmrl/pubs/index.html. Accessed December 13, 2005.

EPA (Environmental Protection Agency). 2001b. Occurrence of Unregulated Contaminants in Public Water Systems — A National Summary. EPA Office of Water. EPA 815-P-00-002. June 2001.
http://www.epa.gov/safewater/ucmr/ucm_rounds_1-2.html. Accessed December 13, 2005.

EPA (Environmental Protection Agency). 2001c. Unregulated Contaminant Monitoring Rule 1 (UCMR 1). Table: UCMR Monitoring List. http://www.epa.gov/safewater/ucmr/ucmr1/factsheet.html#list. Accessed December 13, 2005.

EPA (Environmental Protection Agency). 2003. National Management Measures to Control Nonpoint Source Pollution from Agriculture. EPA 841-B-03-004. July 2003. http://www.epa.gov/owow/nps/agmm/. Accessed December 13, 2005.

EPA (Environmental Protection Agency). 2003. U.S. EPA Toxics Release Inventory — Reporting Year 2003 Public Data Release, Summary of Key Findings. Toxics Release Inventory Program. http://www.epa.gov/tri/tridata/tri03/index.htm. Accessed December 13, 2005.

EPA (Environmental Protection Agency). 2004a. Protecting Water Quality from Agricultural Runoff. EPA 841-F-03-004. Available from http://www.epa.gov/owow/nps/agriculture.html. Accessed December 13, 2005.

EPA (Environmental Protection Agency). 2004b. 2004 Edition of the Drinking Water Standards and Health Advisories. EPA 822-R-04-005. Available from http://www.epa.gov/waterscience/drinking/standards/dwstandards.pdf. Accessed December 13, 2005.

EPA (Environmental Protection Agency). 2005a. List of Contaminants and Their MCLs. http://www.epa.gov/safewater/mcl.html#mcls. Accessed December 13, 2005.

EPA (Environmental Protection Agency). 2005b. Drinking Water Contaminant Candidate List 2. http://www.epa.gov/safewater/ccl/ccl2_list.html. Accessed December 13, 2005.

EPA (Environmental Protection Agency). 2005c. Unregulated Contaminant Monitoring Rule 2 (UCMR 2). UCMR 2 Contaminants and Corresponding Analytical Methods. http://www.epa.gov/safewater/ucmr/ucmr2/factsheet.html#list. Accessed December 13, 2005.

EPA (Environmental Protection Agency). 2005d. PPCPs as Environmental Pollutants. National Exposure Research Laboratory. Environmental Sciences. http://www.epa.gov/esd/chemistry/pharma/new.htm. Accessed December 13, 2005.

FR (Federal Register). 1996. Vol. 61, No. 94. Tuesday, May 14, 1996. Final Rule. National Primary Drinking Water Regulations: Monitoring Requirements for Public

Drinking Water Supplies: Cryptosporidium, Giardia, Viruses, Disinfection Byproducts, Water Treatment Plant Data and Other Information Requirements.

FR (Federal Register). 1998. Vol. 63, No. 24. December 16, 1998. National Primary Drinking Water Regulations: Disinfectants and Disinfection Byproducts; Final Rule.

GAO (U.S. Government Accountability Office). 2005. Chemical Regulation: Options Exist to Improve EPA's Ability to Assess Health Risks and Manage Its Chemical Review Program. GAO-05-458. June 2005.

USGS (U.S. Geological Survey). 2003. Projections of Land Use and Land Cover Change. U.S. Global Change Research Program.
http://www.usgcrp.gov/usgcrp/ProgramElements/recent/landrecent.htm.

USGS (U.S. Geological Survey). 1999. The Quality of Our Nation's Waters. Nutrients and Pesticides. U.S. Geological Survey Circular 122.
http://pubs.usgs.gov/circ/circ1225/html/wq_urban.html.

EPA Guidance for People with Severely Weakened Immune Systems
(co-released with the Centers for Disease Control and Prevention, 1995)
Office of Water (4606) EPA 816-F-99-005 June 1999 www.epa.gov/safewater

Patterson and Kim, The Day America Told the Truth" 1991.
Publisher: Simon & Schuster Published date: 1991

ANNEX

United States
Environmental Protection
Agency

Office of Water (4601M)
Office of Ground Water and Drinking Water
Distribution System Issue Paper

Finished Water Storage Facilities

August 15, 2002

PREPARED FOR:

U.S. Environmental Protection Agency
Office of Ground Water and Drinking Water
Standards and Risk Management Division
1200 Pennsylvania Ave., NW
Washington DC 20004

Prepared by:
AWWA
With assistance from
Economic and Engineering Services, Inc

Background and Disclaimer

The USEPA is revising the Total Coliform Rule (TCR) and is considering new possible distribution system requirements as part of these revisions. As part of this process, the USEPA is publishing a series of issue papers to present available information on topics relevant to possible TCR revisions. This paper was developed as part of that effort.

The objectives of the issue papers are to review the available data, information and research regarding the potential public health risks associated with the distribution system issues, and where relevant identify areas in which additional research may be warranted. The issue papers will serve as background material for EPA, expert and stakeholder discussions. The papers only present available information and do not represent Agency policy. Some of the papers were prepared by parties outside of EPA; EPA does not endorse those papers, but is providing them for information and review.

Additional Information

The paper is available at the TCR web site at:

http://www.epa.gov/safewater/disinfection/tcr/regulation_revisions.html

Questions or comments regarding this paper may be directed to TCR@epa.gov.

Finished Water Storage Facilities

1.0   Introduction

The goal of this document is to review existing literature, research and information on the potential public health implications associated with covered storage reservoirs.

Finished water storage facilities are an important component of the protective distribution system "barrier" that prevents contamination of water as it travels to the customer. Historically, finished water storage facilities have been designed to equalize water demands, reduce pressure fluctuations in the distribution system; and provide reserves for fire fighting, power outages and other emergencies. Many storage facilities have been operated to provide adequate pressure and have been kept full to be better prepared for emergency conditions. This emphasis on hydraulic considerations in past designs has resulted in many storage facilities operating today with larger water storage capacity than is needed for non-emergency usage. Additionally, some storage facilities have been designed such that the high water level is below the hydraulic grade line of the system, making it very difficult to turnover the tank. If the hydraulic grade line of the system drops significantly, very old water may enter the system. If tanks are kept full yet are underutilized, the stored water ages and water quality is affected.

The main categories of finished water storage facilities include ground storage and elevated storage. Finished water storage does not include facilities such as clearwells that are part of treatment or contact time requirements per the Surface Water Treatment Rules. Ground storage tanks or reservoirs can be below ground, partially below ground, or constructed above ground level in the distribution system and may be accompanied by pump stations if not built at elevations providing the required system pressure by gravity. Ground storage reservoirs can be either covered or uncovered. Covered reservoirs may have concrete, structural metal, or flexible covers. The most common types of elevated storage are elevated steel tanks and standpipes. In recent years, elevated tanks supported by a single pedestal have been constructed where aesthetic considerations are an important part of the design process. A standpipe is a tall cylindrical tank normally constructed of steel, although concrete may be used as well. The standpipe functions somewhat as a combination of ground and elevated storage. Only the portion of the storage volume of a standpipe that provides water at or above the required system pressure is considered useful storage for pressure equalization purposes. The lower portion of the storage acts to support the useful storage and to provide a source of emergency water supply. Many standpipes were built with a common inlet and outlet.

2.0   Description of Potential Water Quality Problems

Water quality problems in storage facilities can be classified as microbiological, chemical or physical. Excessive water age in many storage facilities is probably the most important factor related to water quality deterioration. Long detention times, resulting in excessive water age, can be conducive to microbial growth and chemical changes. The excess water age is caused by 1) under utilization (i.e., water is not cycled through the facility), and 2) short circuiting within the reservoir. Poor mixing (including stratification) can exacerbate the water quality problems by

Prepared by AWWA with assistance from Economic and Engineering Services, Inc.

creating zones within the storage facility where water age significantly exceeds the average water age throughout the facility. Distribution systems that contain storage facilities where water cascades from one facility to another (such as pumping up through a series of pressure zones) can result in exceedingly long water age in the most distant tanks and reservoirs. Although the storage facility is normally an enclosed structure, numerous access points can become entry points for debris and contaminants. These pathways may include roof top access hatches and appurtenances, sidewall joints, vent and overflow piping.

Table 1 provides a summary of water quality problems associated with finished water storage facilities.

Table 1
Summary of Water Quality Problems Associated with Finished Water Storage Facilities

Chemical Issues	Biological Issues	Physical Issues
Disinfectant Decay	Microbial Regrowth*	Corrosion
Chemical Contaminants*	Nitrification*	Temperature/Stratification
DBP Formation*	Pathogen Contamination*	Sediment*
Taste and Odors	Tastes and Odors	

*Water quality problem with direct potential health impact.

All issues listed in Table 1 can deteriorate water quality, but only those with direct potential health impacts (identified by an asterisk) are discussed in the following sections or in other White Papers.

2.1 Potential Health Impacts

Various potential health impacts have been associated with the chemical and biological issues identified in Table 1. The Chemical Health Effects Tables (U.S. Environmental Protection Agency, 2002a) provides a summary of potential adverse health effects from high/long-term exposure to hazardous chemicals in drinking water. The Microbial Health Effects Tables (U.S. Environmental Protection Agency, 2002b) provides a summary of potential health effects from exposure to waterborne pathogens.

2.1.1 Sediment

Sediment accumulation occurs within storage facilities due to quiescent conditions which promote particle settling. Potential water quality problems associated with sediment accumulation include increased disinfectant demand, microbial growth, disinfection by-product formation, and increased turbidity within the bulk water. Instances of microbial contamination and disinfection by-product formation due to storage facility sediments are described in the Pathogen Contamination and Microbial Growth section and the Disinfection By-Product formation section, respectively.

2.1.2 Pathogen Contamination and Microbial Growth

Microbial contamination from birds or insects is a major water quality problem in storage tanks. One tank inspection firm that inspects 60 to 75 tanks each year in Missouri and southern Illinois reports that 20 to 25 percent of tanks inspected have serious sanitary defects, and eighty to ninety percent of these tanks have various minor flaws that could lead to sanitary problems (Zelch 2002). Most of these sanitary defects stem from design problems with roof hatch systems and vents that do not provide a watertight seal. Older cathodic protection systems of the hanging type also did not provide a tight seal. When standing inside the tank, daylight can be seen around these fixtures. The gaps allow spiders, bird droppings and other contaminants to enter the tank. Zelch (2002) reports a trend of positive total coliform bacteria occurrences in the fall due to water turnover in tanks. Colder water enters a tank containing warm water, causing the water in the tank to turn over. The warm water that has aged in the tank all summer is discharged to the system and is often suspected as the cause of total coliform occurrences.

Storage facilities have been implicated in several waterborne disease outbreaks in the United States and Europe. In December 1993, a *Salmonella typhimurium* outbreak in Gideon, Missouri resulted from bird contamination in a covered municipal water storage tank (Clark et al. 1996). Pigeon dropping on the tank roof were carried into the tank by wind and rain through a gap in the roof hatch frame (Zelch 2002). Poor distribution system flushing practices led to the complete draining of the tank's contaminated water into the distribution system. As of January 8, 1994, 31 cases of laboratory confirmed salmonellosis had been identified. Seven nursing home residents exhibiting diarrheal illness died, four of whom were confirmed by culture. It was estimated that almost 600 people or 44% of the city's residents were affected by diarrhea in this time period.

A 1993 outbreak of *Campylobacter jejuni* was traced to untreated well water that was likely contaminated in a storage facility that had been cleaned the previous month (Kramer et al. 1996). Fecal coliform bacteria were also detected in the stored water.

In 2000, a City in Massachusetts detected total coliform bacteria in several samples at one of their six finished water storage facilities (Correia, 2002). The tank inspector discovered an open access hatch and other signs of vandalism. This tank was drained and cleaned to remove several inches of accumulated sediment. Three other finished water storage facilities were cleaned in 2001 without being drained and removed from service. The tank closest to the filtration plant was found to contain two to three inches of accumulated sediment and the tanks in outlying areas contained four to six inches of sediment. Shortly after the tanks were returned to service, the City experienced widespread total coliform occurrences in the distribution system (Correia, 2002). The City's immediate response was to boost the free chlorine residual in the distribution system to 4.0 mg/L (including at tank outlets). Also, the distribution system was flushed continuously for two days to remove the contaminated water. These measures resolved the coliform bacteria problem. A boil water order was not required. To prevent the problem from recurring, the City has instituted a tank cleaning program in which all tanks are cleaned on a three year cycle. City engineers are planning to improve water turnover rates by separating the tank inlet and outlet piping.

In 1995, a water district in Maine traced a total coliform bacteria occurrence in the distribution system to two old steel tanks with wooden roofs (Hunt 2002). Upon inspection, many roof shingles were missing and large gaps were present in the tank roofs. After the tanks were

Prepared by AWWA with assistance from Economic and Engineering Services, Inc.

drained, an interior inspection found two feet of accumulated sediment, widespread coating failure on the tank sidewalls, and evidence of human entry. The tanks were cleaned and the distribution system was flushed and disinfected. A boil water order was in place until system water quality was restored. The tanks have since been replaced with a modern preload concrete tank.

Uncovered storage reservoirs provide the greatest opportunity for contaminant entry into the distribution system. These reservoirs are potentially subject to contamination from bird and other animal excrement that can potentially transmit disease-causing organisms to the finished water. Microorganisms can also be introduced into open reservoirs from windblown dust, debris and algae. Algae proliferate in open reservoirs with adequate sunlight and nutrients and impart color, taste and odor to the water on a seasonal basis. Organic matter such as leaves and pollen are also a concern in open reservoirs. Waterfowl are known carriers of many different waterborne pathogens and have the ability to disseminate these pathogens over a wide area. For example, *Vibrio cholerae* has been isolated from feces of 20 species of aquatic birds in Colorado and Utah (Ogg, Ryder and Smith 1989). Waterfowl are known carriers of *S. Montevideo B*, *Vibrio cholerae,* and Hepatitis A virus (Brock 1979) and *E. coli*, Norwalk virus, Coronavirus, Coxsackieviruses, Rotavirus, Astrovirus, and Cryptosporidium (WRc and Public Health Laboratory Service 1997).

Reservoirs with floating covers are susceptible to bacterial contamination and regrowth from untreated water that collects on the cover surface. Birds and animals are attracted to the water surface and may become trapped. Surface water collected on the floating cover of one storage reservoir contained fecal coliform bacteria counts as high as 13,000 per 100 mL and total coliform bacteria counts as high as 33,000 per 100 mL (Kirmeyer et al. 1999). If the cover rips or is otherwise damaged, any untreated water on the cover would mix with the stored water, potentially causing health problems. Floating covers on storage reservoirs are susceptible to rips and tears due to ice damage, vandalism, and/or changing operating water levels.

Based on surveys of professional tank inspection firms, State primacy agencies and utilities, Kirmeyer et al. (1999) concluded that many storage facilities are not being inspected at all. For facilities that are inspected, it is likely that prior to implementation of the Interim Enhanced Surface Water Treatment Rule (IESWTR) they were inspected less frequently than the three-year frequency recommended by AWWA (AWWA Manual M42, 1998). The survey of tank inspection firms indicated that the most frequently documented interval between inspections at that time was six to eight years. Information on inspection practices subsequent to implementation of the IESWTR, which included a prohibition on new uncovered finished water reservoirs, re-focused utility and state regulators on the issues surrounding uncovered reservoirs and floating covers.

The most common problems reported by commercial inspectors in survey responses are: no bug screens on vents and overflows, cathodic protection systems not operating or not adjusted properly, unlocked access hatches, presence of lead paint (interior and exterior), and the presence of paints not approved by NSF International (Kirmeyer et al. 1999). The most common coating problems reported by commercial tank inspectors that relate to water quality (Kirmeyer et al. 1999) are: chemical leaching from incompletely cured coating; corrosion product buildup from excessive interior corrosion; turbidity events during tank filling due to excessive bottom sediment; unknown chemical leaching due to non NSF-61 Coatings; and lead leaching from lead based interior coatings.

The Total Coliform Rule (TCR) was promulgated specifically to identify public water systems that are contaminated or vulnerable to contamination. The total coliform group of organisms is used to indicate the possible presence or absence of pathogens and thus, provides a general indication of whether the water is contaminated. The presence of fecal coliforms or E. coli provides stronger evidence of fecal contamination than does a positive total coliform test and the likely presence of pathogens (Levy et al. 1999). The Total Coliform Rule does not specifically require monitoring at storage reservoirs, however, state primacy agencies have oversight of utility monitoring plans and may require selection of sample sites, such as reservoirs, when appropriate in TCR Monitoring Plans.

The Surface Water Treatment Rule establishes maximum contaminant level goals (MCLGs) for viruses, *Legionella*, HPC, and *Giardia lamblia*. It also includes treatment technique requirements for filtered and unfiltered systems that are specifically designed to protect against the adverse health effects of exposure to these microbial pathogens. The Surface Water Treatment Rule requires that a "detectable" disinfectant residual (or heterotrophic plate count (HPC) measurements not exceeding 500/mL) be maintained in at least 95% of samples collected throughout the distribution system on a monthly basis. A system that fails to comply with this requirement for any two consecutive months is in violation of the treatment technique requirement. Public water systems must monitor for the presence of a disinfectant residual (or HPC levels) at the same frequency and locations as total coliform measurements taken pursuant to the total coliform regulation described above.

The loss of disinfectant residual within a storage facility does not necessarily pose a direct public health threat (many systems throughout the world are operated without use of a disinfectant residual). However, disinfectant decay can contribute to microbiological problems such as growth of organisms within the bulk water or sediment. The rate of decay can be affected by external contamination, temperature, nitrification, exposure to ultraviolet light (sun), and amount and type of chlorine demanding compounds present such as organics and inorganics. Chlorine decay in storage facilities can normally be attributed to bulk water decay rather than wall effects due to the large volume-to-surface area ratio.

A long detention time can allow the disinfectant residual to be completely depleted thereby not protecting the finished water from additional microbial contaminants that may be present in the distribution system downstream of the storage facility. This problem is illustrated in a recent investigation of storage tanks in a large North American water utility's distribution system (Gauthier et al. 2000). An estimation of stored water turnover rate using routine water quality data and hydraulic modeling results found that one tank had a turnover rate of 5.6 to 7.6 days which was probably responsible for the periodic loss of disinfectant residual in the surrounding distribution system and several occurrences of total coliform bacteria. The high residence time was caused by the hydraulic arrangement of the tank and pumping system where most water was pumped directly to consumers and the remaining water was fed to the tank. Water leaving the tank typically had a chlorine residual of 0.05 mg/L.

A detailed discussion of potential health issues associated with microbial growth and biofilms is provided in a separate White Paper.

2.1.3 Nitrification

Nitrification is a potential health concern in finished water storage facilities due to the formation of nitrite and nitrate. Nitrification may occur within storage facilities due to long hydraulic residence times. Under the Safe Drinking Water Act (SDWA), primary MCLs have been established for nitrite-N, nitrate-N, and the sum of nitrite-N plus nitrate-N. The MCLs are 1 mg/L for nitrite-N, 10 mg/L for nitrate-N, and 10 mg/L for nitrite + nitrate (as N). The nitrite and nitrate MCLs are applicable at the point-of-entry to the distribution system, not within the distribution system where nitrification is most likely to occur. Review of nitrification episodes and information gathered from the literature indicates that an MCL exceedence within the distribution system due to nitrification is unlikely, unless source water nitrate-N or nitrite-N levels are close to their applicable MCLs. Potential public health issues associated with nitrification are discussed in the Nitrification White Paper.

2.1.4 Chemical Contaminants

Coating materials are used to prevent corrosion of steel storage tanks and to prevent moisture migration in concrete tanks. Through the 1970's, coatings used in finished water storage facilities were primarily selected because of their corrosion resistance and ease of application. This led to the use of industrial products like coal tars, greases, waxes and lead paints as interior tank coatings. These products offered exceptional corrosion performance but unknowingly contributed significant toxic chemicals to the drinking water. Grease coatings can differ greatly in their composition from vegetable to petroleum based substances and can provide a good food source for bacteria, resulting in reduced chlorine residuals and objectionable tastes and odors in the finished water (Kirmeyer et al. 1999).

An old grease coating on a storage tank interior in the state of Florida was suspected of causing water quality problems in the distribution system such as taste and odor, high chlorine requirements and a black slime at the customers tap. The Wisconsin Avenue 500,000 gallon elevated tank was originally coated with a petroleum grease coating when it was built in 1925. In 1988, the storage facility was cleaned and the grease coating was reapplied. In 1993, a tank inspection revealed that the grease had sagged off the tank walls and deposited a thick accumulation of black loose ooze in the bottom bowl of the tank (6-8 inches deep). A thin film of grease continued to coat the upper shell surfaces. Although this material had performed well as a corrosion inhibitor, it was introducing debris into the distribution system as well as creating a possible food source and environment for bacteria. The City decided to completely remove the grease and reapply a polyamide epoxy system. This work was completed in 1996 (Kirmeyer et al. 1999). Since the tank was returned to service, water quality has markedly improved. The required chlorine dosage rate has decreased from 4.0-5.0 mg/L to 3.5 mg/L. The chlorine residual at the tank outlet has improved from <1.0 mg/L to 1.4 mg/L. No more "black slime" complaints have been received.

The East Bay Municipal Utility District used hot-mopped coal tar as the standard interior coating system for tanks through the 1960's then discontinued its use due to concerns over VOCs (Irias, 2000). When manufacturer's directions and AWWA standards are not followed correctly, these coatings can leach organics into the finished water. Volatile organic compounds could be introduced to the stored water if sufficient curing time is not allowed after coating application (Kirmeyer et al. 1999). Burlingame and Anselme (1995) cite examples of odiferous organic solvents leaching from reservoir linings. Elevated levels of alkyl benzenes and polycyclic aromatic hydrocarbons (PAHs) have been reported in reservoirs with new bituminous coatings and linings (Yoo et al, 1984; Krasner and Means, 1985; Alben, 1980).

Alben et al (1989) studied leaching of organic contaminants from flat steel panels lined with various coatings, including vinyl, chlorinated rubber, epoxy, asphalt, and coal tar. Emphasis was given to the rate of leachate production and leachate composition. The test water was GAC processed tap water with a pH of 8 to 9. Leaching rates (mg/m^2-day or ug/L-day) were assessed over a period of 30 days. Organic contaminants were found at parts-per-billion levels in water compared to parts-per-thousand levels in the coating. Detailed findings of the leaching study are provided in the Permeation and Leaching White Paper.

Solvents, adhesives and other materials used to repair floating covers could potentially contaminate the drinking water as storage reservoirs are not always drained to accomplish the repair. For example, Philadelphia formerly used trichloroethylene as a solvent to clean areas to be repaired on a Hypalon cover prior to making repairs (Kirmeyer et al. 1999). In 1984, Philadelphia repaired one basin's Hypalon cover 200 times. This Hypalon cover has since been replaced with a more durable, polypropylene cover.

Improper installation procedures may result in worker and public exposure to chemicals. For example, odor complaints at a Duval County, Florida utility led to a discovery of ethyl benzene contamination of the distribution system water (Carter, Cohen, and Hilliard, 2001). The source of ethyl benzene was determined to be a polyamide solvent applied to a ground storage water tank prior to painting. It is likely that solvent vapors carried over to an adjacent on-line aeration tower and became dissolved in the water. Flushing was conducted immediately as a response to this incident, and no samples were analyzed prior to flushing. After flushing, a distribution system water sample contained 0.004 mg/L ethyl benzene. The MCL for ethyl benzene is 0.7 mg/L and the secondary standard MCL threshold for odor is 0.03 mg/L.

When volatile compounds have entered a water distribution system through source contamination or contamination within the distribution system, storage facilities with a free water surface and reservoir vents can serve as a pathway for volatilization to the atmosphere. Walski (1999) describes an analysis method for estimating the loss of volatiles at a storage facility.

The National Sanitation Foundation (NSF) International and Underwriters Laboratory (UL) certify coatings and other products against ANSI/NSF Standard 61 (NSF 1996b), a nationally accepted standard addressing the health effects of water contact materials. Details on the NSF certification procedure are provided in the Permeation and Leaching White Paper.

The following AWWA standards were developed to ensure that approved coatings function as intended:

- D102 Coating Steel Water Storage Tanks
- D104 Cathodic Protection for Interior of Steel Water Tanks
- D110 Wire- and Strand-Wound Circular Prestressed-Concrete Water Tanks
- D130 Flexible Membrane Lining and Floating Cover Materials for Potable Water Storage

Twenty-one volatile organic compounds (VOCs) and 33 synthetic organic compounds (SOCs) are currently regulated under the Safe Drinking Water Act Phase I, II, and V Rules based on health effects that may result from long-term exposures. Compliance is determined based on annual average exposure measured at the point of entry to the distribution system.

2.1.5   Disinfection By-Products

Storage facilities provide opportunities for increased hydraulic residence times, allowing more time for disinfection by-products (DBPs) to form. Rechlorination within storage facilities exposes the water to higher chlorine dosages, potentially increasing disinfection by-product formation. Higher water temperatures in steel tanks during summer seasons can increase disinfection by-products as the chemical reactions proceed faster and go further at higher temperatures. Storage facilities with new interior concrete surfaces often have elevated pH levels that can also increase trihalomethane formation.

The USEPA has identified the following potential adverse health effects associated with HAA5 and TTHMs:

"Some people who drink water containing haloacetic acids in excess of the MCL over many years may have an increased risk of getting cancer. Some people who drink water containing trihalomethanes in excess of the MCL over many years may experience problems with their liver, kidneys, or central nervous system, and may have an increased risk of getting cancer."

The forthcoming Stage 2 Disinfectants and Disinfection By-Products Rule is expected to include a new monitoring and reporting approach. Compliance with TTHMs and HAA5 standards will be based on a locational running annual average using monitoring data gathered at new monitoring locations selected to capture representative high levels of occurrence. MCL violations could potentially occur at single locations such as a finished water storage facility due to site-specific situations, including excessive water age or chlorine addition at the storage facility.

3.0 Prevention/Mitigation Methods

3.1 Indicators of Water Quality Problems within Storage Facilities

There are several indicators that may suggest water quality problems are occurring within storage facilities. These include aesthetic considerations that may be identified by consumers, as well as the results of storage facility monitoring efforts. It should be noted that indicators can be triggered by factors other than water age, such as insufficient source water treatment, pipe materials, and condition/age of distribution system and storage facility.

Aesthetic Indicators

The following indicators may be identified during water consumption:

- Poor taste and odor – Aged, stale water provides an environment conducive to the growth and formation of taste and odor causing microorganisms and substances. Improperly cured coatings can impart taste and odor to the stored water.
- Sediment accumulation – Improperly applied coatings can slough off reservoirs and accumulate at the bottom. Sediment carried into the storage facility from the bulk water can accumulate within the reservoir if reservoir maintenance and cleaning are not routinely performed.
- Water temperature – Stagnant water will approach the ambient temperature. Temperature stratification within reservoirs will impede mixing. Turnover due to stratification can entrain accumulated sediment.

Monitoring Indicators

The following indicators require sample collection and analysis:

- Depressed disinfectant residual – Chlorine and chloramines undergo decay over time.
- Elevated DBP levels – The reaction between disinfectants and organic precursors occur over long periods.
- Elevated bacterial counts (i.e., heterotrophic plate count).
- Elevated nitrite/nitrate levels (nitrification) for chloraminating systems.

3.2 Water Quality Monitoring and Modeling

Water quality monitoring and modeling are useful tools to assess the impact storage may be having on water quality in a distribution system. Studies can be conducted to define current or potential water quality problems in storage facilities. Water quality monitoring at storage facilities is not required by any specific federal regulations.

Monitoring within a storage facility can supplement tank inlet or outlet monitoring where short-circuiting or lack of use may cause water quality to vary widely within the tank. When detailed investigation of a storage facility's impact is warranted, the ideal sampling program would capture water quality conditions throughout the storage facility, both vertically and spatially. Kirmeyer et al. (1999) recommended the following monitoring parameters: free and total chlorine residual, temperature, HPC, total and fecal coliform bacteria, pH, turbidity, and total dissolved solids. Monitoring in storage facilities can often be a difficult task and can present a safety issue because sampling taps or access ports are often not installed during the initial construction and utility workers must generally climb the tank and collect grab samples through the roof access hatchways.

Direct monitoring may not detect all potential water quality problems. For example, tank effluent sampling can result in zero bacteria counts, but microorganisms can still be present as biofilms on tank surfaces, in tank sediment or in the water (Smith and Burlingame 1994).

According to Grayman and Kirmeyer (2002), modeling can provide information on what will happen in an existing, modified, or proposed facility under a range of operating situations. There are two primary types of models: physical scale models and mathematical models. Physical scale models are constructed from materials such as wood or plastic. Dyes or chemicals are used to trace the movement of water through the model. In mathematical models, equations are written to simulate the behavior of water in a tank or reservoir. These models range from detailed representations of the hydraulic mixing phenomena in the facility called computational fluid dynamics (CFD) models to simplified conceptual representations of the mixing behavior called systems models. Information collected during monitoring studies can be used to calibrate and confirm both types of models.

3.3   Tank Inspections

Like water quality monitoring, tank inspections provide information used to identify and evaluate current and potential water quality problems. Both interior and exterior inspections are employed to assure the tank's physical integrity, security, and high water quality. Inspection type and frequency are driven by many factors specific to each storage facility, including its type (i.e. standpipe, ground tank, etc), vandalism potential, age, condition, cleaning program or maintenance history, water quality history, funding, staffing, and other utility criteria. AWWA Manual M42, Steel Water Storage Tanks (1998) provides information regarding inspection during tank construction and periodic operator inspection of existing steel tanks. Specific guidance on the inspection of concrete tanks was not found in the literature. However, the former AWWA Standard D101 document may be used as a guide to inspect all appurtenances on concrete tanks. Concrete condition assessments should be performed with guidance from the tank manufacturer. Soft, low alkalinity, low pH waters may dissolve the cementitious materials in a concrete reservoir causing a rough surface and exposing the sand and gravel. The concern is that in extreme cases, the integrity of reinforcing bars may be compromised. Sand may collect on the bottom of the storage facility during this process.

Routine inspections typically monitor the exterior of the storage facility and grounds for evidence of intrusion, vandalism, coating failures, security, and operational readiness. Based on a literature review and project survey, Kirmeyer et al. (1999) suggested that routine inspections be conducted on a daily to weekly basis. Where SCADA systems include electronic surveillance systems, alarm conditions may substitute for physical inspection.

Periodic inspections are designed to review areas of the storage facility not normally accessible from the ground and hence not evaluated by the routine inspections. These inspections usually require climbing the tank. Periodic inspections, like routine inspections, are principally a visual inspection of tank integrity and operational readiness. Based on a literature review and project survey, Kirmeyer et al. (1999) suggested that periodic inspections be conducted every 1 to 4 months.

Comprehensive inspections are performed to evaluate the current condition of storage facility components. These inspections often require the facility to be removed from service and drained unless robotic devices or divers are used. The need for comprehensive inspections is generally recognized by the water industry. AWWA Manual M42 (1998) recommends that tanks be drained and inspected at least once every 3 years or as required by state regulatory agencies. Most states do not recommend inspection frequencies thereby leaving it to the discretion of the utility. States that do have recommendations are Alabama (5 years), Arkansas (2 years), Missouri (5 years), New Hampshire (5 years), Ohio (5 years), Rhode Island (external once per year; internal, every five years), Texas (annually), and Wisconsin (5 years). Kirmeyer et al. (1999) recommend that comprehensive inspections be conducted every 3 to 5 years for structural condition and possibly more often for water quality purposes.

Uncovered finished water reservoirs have unique problems. Consequently, water utilities have ceased constructing such facilities. As noted previously, the IESWTR prohibits construction of new uncovered finished water reservoirs in the U.S. Under the LT2ESWTR, existing uncovered finished water reservoirs will be managed in accordance with a state approved plan, if the facility is not covered subsequent to the rule's implementation. Flexible membrane covers are one means of enclosing uncovered reservoirs and these types of facilities also require specific routine, periodic, and comprehensive inspections to ensure the cover's integrity.

3.4   Maintenance Activities

Storage facility maintenance activities include cleaning, painting, and repair to structures to maintain serviceability. Based on a utility survey conducted by Kirmeyer et al. (1999), it appears that most utilities that have regular tank cleaning programs employ a cleaning interval of 2 to 5 years. This survey also showed that most tanks are painted (exterior coating) on an interval of 10 to 15 years.

The following existing standards are relevant to disinfection procedures and approval of coatings:

- ANSI/NSF Standard 61, and
- Ten States Standards (Great Lakes...1997)
- AWWA Manuals

- AWWA M25 – Flexible-Membrane Covers and Linings for Potable-Water Reservoirs (1996)

- AWWA M42 – Steel Water-Storage Tanks (1998)

• AWWA Standards

- AWWA Standard C652-92 Disinfection of Storage Facilities (AWWA 1992) provides guidance for disinfection when returning a storage facility to service.

- AWWA Standard D102 recognizes general types of interior coating systems including:
  ➢ Epoxy,
  ➢ Vinyl,
  ➢ Enamel, and
  ➢ Coal-Tar

Each of the coating systems listed under AWWA Standard D102 has provided satisfactory service when correctly applied (AWWA 1998). Other coating systems have been successfully used including chlorinated rubber, plural-component urethanes, and metalizing with anodic material (AWWA 1998). Epoxy and solvent-less polyurethanes interior coating systems are most likely to meet strict environmental guidelines and AWWA and NSF Standards (Jacobs 2000). Spray metalizing using zinc, aluminum or a combination of both is also a promising alternative. Coal tar coating systems are not common in eastern U.S. as the coatings installed in the 1950s and 1960s have mostly been replaced or the tanks themselves have been removed from service. Coal tar is still in use in California where it is often applied over an epoxy system on tank floors (Lund, 2002).

ANSI/NSF 61 (National Sanitation Foundation 1996) is a nationally accepted standard that protects stored water from contamination via products which come into contact with water. Products covered by NSF 61 include pipes and piping appurtenances, nonmetallic potable water materials, coatings, joining and sealing materials (i.e. gaskets, adhesives, lubricants), mechanical devices (i.e. water meters, valves, filters), and mechanical plumbing devices. NSF 61 was reviewed and certified by the American National Institute of Standards (ANSI) which permitted the use of the standard by other independent testing agencies such as Underwriters Laboratories. With the development of this ANSI/NSF-61 Standard, the approval and reporting for tank coatings process is now standardized. State agencies that previously had independent coating approval programs discontinued these programs and adopted the ANSI/NSF 61 Standard. Details on the ANSI/NSF 61 certification procedure are provided in the Permeation and Leaching White Paper.

Coating manufacturers provide technical specifications for proper coating application and curing. Utilities or their consulting engineer provide technical specifications and drawings describing the specific project. Trained and certified coating inspectors provide quality control during coating application. The National Association of Corrosion Engineers has a certification program for coating inspectors.

Kirmeyer et al. (1999) recommended that covered facilities be cleaned every three to five years, or more often based on inspections and water quality monitoring, and that uncovered storage facilities be cleaned once or twice per year. Commercial diving contractors can be used to clean and inspect storage facilities that cannot be removed from service. AWWA Standard C652-92 provides guidelines for disinfection of all equipment used to clean storage facilities.

Three finished water steel elevated spheroids at the City of Brookfield Water Utility in Brookfield, Wisconsin were the subject of a field study (Kirmeyer et al. 1999) conducted to document the underwater cleaning process and its water quality impacts. The time since last cleaning was 15 years for one tank and 7 years for the other two tanks. The tank with the longest cleaning interval contained the most accumulated sediment (28 inches maximum depth compared to 4-12 inches in the other two tanks), and the highest HPC bacteria levels before cleaning (1300/mL compared to 640 and 80/mL in the other two tanks). As a result of underwater cleaning, HPC bacteria and turbidity levels were significantly reduced.

Maintenance of the cathodic protection system is a component of controlling corrosion and degradation of the submerged coated surface of finished water storage facilities. AWWA Standard D104 (AWWA 1991) provides guidelines on system inspection and maintenance.

3.5   Operations Activities

As noted previously, water age is an important variable in managing water quality in finished water storage. Operationally, water age in these facilities is managed by routine turn over of the stored water and fluctuation of the water levels in storage facilities. Kirmeyer et al. (1999) recommended a 3 to 5 day complete water turnover as a starting point, but cautioned that each storage facility be evaluated individually and given its own turnover goal. Water storage management for water quality must take into account influent water quality, environmental conditions, retention of fire flow, and demand management, as well as factors specific to the design and operation of the tank such as velocity of influent water, operational level changes, and tank design. Consequently, water level fluctuations in a distribution system are managed as an integrated operation within pressure zones, demand service areas, and the system as a whole rather than on an individual tank basis. Available guidelines for water turnover rates are summarized in Table 2.

From a field perspective, the Philadelphia Water Department estimated mean residence time and turnover rate in several standpipes by measuring fluoride residual and water levels. Mean residence time of water in the standpipes was determined to be 50 percent longer than expected because "old" water re-entered the standpipes from the distribution system. One major conclusion from this work was that for water to get out into the distribution system and away from the standpipes, standpipe drawdown needs to correspond to peak demands or precede peak demands. (Burlingame, Korntreger and Lahann 1995).

Philadelphia also demonstrated how operational changes can reduce the hydraulic detention time needed to restore or maintain a disinfectant residual within the storage facility. During normal operation, the water levels in the storage facilities were allowed to drop an additional ten feet in elevation, decreasing the mean residence time by two to three days. As a result, disinfectant residuals were maintained at acceptable levels, even during the summer months (Burlingame and Brock 1985).

Table 2
Guidelines on Water Turnover Rate

Source	Guideline	Comments
Georgia Environmental Protection Division	Daily turnover goal equals 50% of storage facility volume; minimum desired turnover equals 30% of storage facility volume	As part of this project, state regulators were interviewed by telephone.
Virginia Department of Health, Water Supply Engineering Division, Richmond, VA	Complete turnover recommended every 72 hours	As part of this project, state regulators were interviewed by telephone.
Ohio EPA	Required daily turnover of 20%; recommended daily turnover of 25%	Code of state regulations; turnover should occur in one continuous period rather than periodic water level drops throughout the day.
Baur and Eisenbart 1988	Maximum 5 to 7 day turnover	German source, guideline for reservoirs with cement-based internal surface.
Braid 1994	50% reduction of water depth during a 24 hour cycle	Scottish source.
Houlmann 1992	Maximum 1 to 3 day turnover	Swiss source.

Source: Kirmeyer et al. (1999)

The Greater Vancouver Water District (GVWD) completed a field study of operational changes and their effects on stored water quality (Kirmeyer et al. 1999). Historical water quality monitoring indicated that finished water reservoirs often had chlorine residuals below 0.2 mg/L and HPC levels above 500 cfu/mL. At the Central Park Reservoir, chlorine residuals were low or non-detectable, and HPC levels were >10,000 cfu/mL. Operational practices for the Central Park Reservoir and the Vancouver Heights Reservoirs had resulted in time periods when the water would remain stagnant, with little or no exchange with water from the supply main (daily turnover rate between 0 and 10 percent). At the Vancouver Heights Reservoir, the average daily turnover was increased from 10% to more than 100% changing the reservoir from one that floated on the system to a flow through operation. Operational changes made at the Central Park Reservoir improved daily water turnover rate to 50 percent. Monitoring after these operational changes indicated that chlorine residual levels were above 0.2 mg/L and HPC bacteria counts were consistently less than 500 cfu per mL.

The Consumers New Jersey Water Company experimented with a new standpipe to improve its water turnover rate (Kirmeyer et al. 1999). This facility was underutilized and had a turnover rate greater than 8 days. Control of the booster pumps feeding the service area was changed from an older elevated tank's operating level to the new standpipe's operating level. Various operating water level ranges were tested under both summer and winter demand conditions. During the summer study period, the operational changes did not increase chlorine residuals in the new standpipe. It appeared that the newer water was being pumped directly to the customers, and the older water was being returned to the storage facilities. During the winter study period, chlorine residuals in the new standpipe increased minimally from 0.1 to 0.2 mg/L. The turnover rate was reduced from 8.3 days to 4.6 days. Equipment problems were encountered as a result of longer pumping periods. When the standpipe's operating range was changed, the booster pump feeding it cycled 1.5 times per day for a longer period instead of 6 times per day. The booster pump motor became overheated and failed, causing damage to the pump starter and the main breaker. This field study illustrates that operational changes are not necessarily straightforward, and that water quality testing is important for evaluating proposed changes. To further improve mixing effects in new storage facilities, Consumers New Jersey is now using separate inlet and outlet piping arrangements.

The Eugene Water and Electric Board (EWEB) in Oregon had a difficult time maintaining chlorine residuals in their upper service levels, primarily due to chlorine decay in the bulk water over extended time periods. The EWEB operations staff determined that improved chlorine residuals could possibly be realized by changes in pump control. Historically, the pump station that fed each upper level reservoir pumped independently of each pump station in the service levels below it. By synchronizing pump station operations, EWEB found that water could be moved from the first level service area directly to any of the upper level service areas without first being discharged to an intermediate service level reservoir. The new pumping scheme decreased the water's residence time in the intermediate reservoirs and improved chlorine residuals throughout the upper service level storage and distribution system. Chlorine residual was not detectable in the upper level reservoirs before the operating change. Afterwards, the chlorine residual in the upper level reservoirs ranged from 0.1 to 0.4 mg/L.

Mixing processes within a storage facility should be controlled to minimize water age (Grayman et al. 2000). When mixing does not occur throughout the storage facility, stagnant zones can form where water age will exceed the overall average water age in the facility. Therefore, mixed flow is preferable to plug flow in distribution system storage. Mixing can be encouraged through the development of a turbulent jet. Mixing a fluid requires a source of energy input, and in a storage facility, this energy is normally introduced from the facility's inflow. As the water enters the facility, jet flow occurs and the ambient water is entrained into the jet and circulation patterns are formed, resulting in mixing. In order to have efficient mixing, the jet flow must be turbulent, and its path must be long enough to allow for the mixing process to develop. In order to assure turbulent jet flow, the following relationship between inflow (Q, in gallons per minute) and inlet diameter (d, in feet) must hold:

$$Q/d > 11.5 \text{ at } 20° \text{ C}$$
$$Q/d > 17.3 \text{ at } 5° \text{ C}$$

Temperature differences between the inflow and the ambient water temperature within the storage facility can cause the water to form stratified layers that do not mix together. Stratification is more common in tall tanks such as standpipes and tanks with large diameter inlets. It can be avoided by increasing the inflow rate. The critical temperature difference, $\cong T$ in °C which can lead to stratification, can be estimated based on the following equation:

$$\cong T = C \, Q^2/(d^3 H^2)$$

Where:

Prepared by AWWA with assistance from Economic and Engineering Services, Inc.

150

C is a coefficient depending on inlet configuration, buoyancy type, and tank diameter
Q = inflow rate (cfs)
H = depth of water (feet)
d = inlet diameter (feet)

Booster disinfection may be required to restore disinfectant residuals at a storage facility. Either a continuous rechlorination system or a batch system can be employed depending on the need. Batch chlorination is used to restore the chlorine residual, to disinfect an existing biological population, or to destroy a taste and odor condition. Free chlorine is the most common secondary disinfectant. Disinfectant can be added at the reservoir inlet, outlet, or within the storage facility if it is equipped with a system to enhance circulation. Chlorine addition at the outlet is normally preferred over the inlet unless the residual is nearly depleted when entering the facility. Conventional rechlorination stations, whether controlled by on/off, flow pacing, or chlorine residual pacing may create a chlorine residual of unpredictable levels. Due to the dynamic nature of flow and chlorine demand in most water distribution systems, these methods of rechlorination can lead to periodic over- and under-feeding. Where rechlorination is in use, careful consideration must be given to storage facility operations. For example, seasonal changes in water demand and temperatures can directly impact rechlorination practices.

Operation of the rechlorination system must also consider the impacts on additional formation of disinfection by-products. A utility practicing chloramination for disinfection must carefully evaluate and monitor any rechlorination process. The mixing of free chlorine with chloramines can result in the loss of free chlorine residual if not conducted properly. If done correctly, chloramine levels can be increased with the addition of chlorine, depending on the level of residual ammonia present. If ammonia concentrations are insufficient, ammonia addition prior to chlorine addition may be required. Additional information related to rechlorination and blending of chlorinated and chloraminated waters is provided in the Nitrification White Paper. Batch chlorination can be accomplished by chlorine injection at the inlet pipe or by chlorine addition into the storage facility contents through hatches or a recirculation system.

Management of distribution systems requires appropriate skills and training. Public water systems employ systems operators that are properly trained and certified per EPA's operator certification guidelines (EPA 1999). These guidelines, required as part of the 1996 Amendments to the Safe Drinking Water Act, provide States with the minimum standards for developing, implementing and enforcing operator certification programs. The guidelines help to ensure that distribution systems, including finished water storage facilities, are operated in a proper manner.

3.6   Design of Storage Facilities

The sizing, number, and type of storage facilities affect a water system's ability to manage water quality while providing an adequate water supply with adequate pressure. Capital planning necessitates installation of facilities that have excess capacity for water storage and distribution. Standard design guidelines for hydraulic considerations in the planning and construction of tanks are available in:

- AWWA Manual M32 Distribution Network Analysis for Water Utilities (AWWA 1989)
- Modeling, Analysis and Design of Water Distribution Systems (AWWA 1995c)
- Hydraulic Design of Water Distribution Storage Tanks (Walski 2000)

These guidelines ensure adequate fire flow to meet applicable codes and rating systems as well as hydraulics of water storage. State regulations address design features related to tank sizing, siting, penetrations, coatings and linings through reference to industry recognized codes and manuals (i.e. AWWA, NSF International and 10 States Standards). A discussion relating fire flow requirements to storage volume and water age is provided in the Water Age White Paper. Findings suggest that volumetric increases are site-specific and cannot be generalized.

Design guidelines addressing water quality include:

- *Maintaining Water Quality in Finished Water Storage Facilities* (Kirmeyer et al. 1999)
- *Water Quality Modeling of Distribution System Storage Facilities* (Grayman et al. 2000)

Appurtenances on storage facilities, such as vents, hatches, drains, wash out piping, sampling taps, overflows, valves, catwalk, etc., can be critical to maintaining water quality. The Ten State Standards (Great Lakes…1997) provides recommended design practices for appurtenances.

Design considerations include mixing to preclude dead zones and to maintain a disinfectant residual. Guidelines for momentum-based mixing can be found in Grayman et al. (2000). Other types of mixings systems are described in Kirmeyer et al. (1999).

4.0 Summary

Microbiological, chemical, and physical water quality problems can occur in finished water reservoirs that are under-utilized or poorly mixed. Poor mixing can be a result of design and/or operational practices. Several guidance manuals have been developed to address design, operations, and maintenance of finished water reservoirs. Water quality issues that have the potential for impacting public health include DBP formation, nitrification, pathogen contamination, and increases in VOC/SOC concentrations. Elevated DBP levels within storage facilities could result in an MCL violation under the proposed Stage 2 Disinfectants and Disinfection Byproduct Rule, based on a locational running annual average approach. A separate White Paper on Nitrification indicates that nitrite and/or nitrate levels are unlikely to approach MCL concentrations within the distribution system due to nitrification unless finished water nitrate/nitrite levels are near their respective MCLs. Pathogen contamination from floating covers or unprotected hatches is possible. Recommended tank cleaning and inspection procedures have been developed by AWWA and AWWARF to address these issues. Elevated levels of VOCs and SOCs have been measured in finished water storage facilities. AWWA and NSF standards have been developed to ensure that approved coatings function as intended. Addition data and evaluation would be required to determine if there is a significant potential for coatings and other products used in distribution system construction and maintenance to cause an

MCL violation based on sampling within the distribution system rather than the currently required monitoring at the point of entry.

5.0 Secondary Considerations

5.1 Water Disposal Issues

When storage facilities are drained prior to cleaning or inspection, the water must be disposed of in accordance with local and State regulations. If the water contains a chlorine residual, dechlorination may be required. The National Pollution Discharge Elimination System is a Federal program established under the Clean Water Act, aimed at protecting the nation's waterways from point and non-point sources of pollution. Effluent limitations vary depending on receiving water characteristics (use classification, water quality standards, and flow characteristics) and discharge characteristics (flow, duration, frequency). Failure to comply with state regulations for such releases can result in legal action against the water utility including monetary compensation and punitive fines. AWWA Standard C652 describes dechlorination procedures for storage tanks.

5.2 Safety Issues

Safety is addressed primarily through referencing Occupational Safety and Health Administration (OSHA) regulations. OSHA regulations address confined space issues (entering the tanks), climbing the tanks, removing lead from tanks, and repainting. The 1994 OSHA Compliance Directive for the Interim Final Rule on Lead Exposure in Construction (29 CFR 1926.62) requires worker protective gear, compliance plans and monitoring equipment for any removal of lead-based coatings. OSHA's Fall Protection Standard applies to all construction sites where workers risk a fall of six feet or more. OSHA's Confined Space Rule (29 CFR 1910.146) requires employers to have written programs and permits for employees working in confined spaces including storage facilities.

EPA Title 10 regulations, issued in 1994, require training and certification of workers handling certain lead bearing materials. Special equipment has been developed for removing lead-based coatings, including vacuum blasting and power tools, and portable mini-containments for reservoirs and standpipes. The Structural Steel Painting Council (SSPC) has developed standards for containment equipment and for power and hand tools used in this application.

Bibliography

Alben, K., A. Bruchet, and E. Shpirt. 1989. *Leachate From Organic Coating Materials Used in Potable Water Distribution Systems*. Denver, Colo.: AWWA and AwwaRF.

Atkinson, R. 1995. A Matter of Public Health: Contamination of Water Storage Tanks. *Missouri Municipal Review*. June Issue.

AWWA (American Water Works Association). 1987a. *AWWA Manual M25—Flexible Membrane Covers and Linings for Potable-Water Reservoirs*. Denver, Colo.: AWWA.

AWWA (American Water Works Association). 1987b. *AWWA Standard for Factory-Coated Bolted Steel Tanks for Water Storage*. AWWA D103-87. Denver, Colo.: AWWA.

AWWA (American Water Works Association). 1989. *AWWA Manual M32—Distribution Network Analysis for Water Utilities*. Denver, Colo.: AWWA.

AWWA (American Water Works Association). 1991. *AWWA Standard for Automatically Controlled, Impressed-Current Cathodic Protection for the Interior of Steel Water Tanks. AWWA D104-91. Denver, Colo.: AWWA.*

AWWA (American Water Works Association). 1992. *AWWA Standard for Disinfection of Water Storage Facilities*. AWWA C652. Denver, Colo.: AWWA.

AWWA (American Water Works Association). 1995a. *AWWA Standard for Circular Prestressed Concrete Water Tanks with Circumferential Tendons*, AWWA D115-95. Denver, Colo: AWWA.

AWWA (American Water Works Association). 1995b. *AWWA Standard for Wire- and Strand-Wound Circular Prestressed-Concrete Water Tanks*. AWWA D110-95. Denver, Colo: AWWA.

AWWA (American Water Works Association). 1995c. *Modeling, Analysis, and Design of Water Distribution Systems*. Denver, Colo: AWWA.

AWWA (American Water Works Association). 1995d. *Water Transmission and Distribution*. Second edition. Denver, Colo.: AWWA.

AWWA (American Water Works Association). 1996a. *AWWA Standard for Flexible-Membrane Lining and Floating-Cover Materials for Potable-Water Storage*. AWWA D130. Denver, Colo.: AWWA.

AWWA (American Water Works Association). 1996. *AWWA Standard for Welded Steel Tanks for Water Storage*. AWWA D100-96. Denver, Colo.: AWWA.

AWWA (American Water Works Association). 1997. *AWWA Standard for Coating Steel Water Storage Tanks*. AWWA D102-97. Denver, Colo.: AWWA.
AWWA (American Water Works Association). 1998. *AWWA Manual M42—Steel Water-Storage Tanks*. Denver, Colo.: AWWA.

Barsotti, M.G., J. Fay, R. Pratt, E. George, and E. Hansen. 2000. Investigating and Controlling HAA$_5$ Levels Within a Complex Water Transmission and Storage System. In *Proc. Of the 1994 AWWA Water Quality and Technology Conference*. Denver, Colo.: AWWA.

Beuhler, M.D., D.A. Foust, and R.W. Mann. 1994. Monitoring to Identify Causative Factors of Degradation of Water Quality: What to Look for. In *Proc. Of the 1994 Annual AWWA Conference*. Denver, Colo.: AWWA.

Brock, T.D., ed. 1979. *Biology of Microorganisms*. 3rd ed. Englewood Cliffs, NJ: Prentice-Hall, Inc.

Burlingame, G.A., and G.L. Brock. 1985. Water Quality Deterioration in Treated-Water Storage Tanks. In *Proc. of the 1985 Annual AWWA Conference*. Denver, Colo.: AWWA.

Burlingame, G.A., G. Korntreger, and C. Lahann. 1995. The Configuration of Standpipes in Distribution Affects Operations and Water Quality. *Jour. NEWWA*, 95(12):218-289.

Carter, C., M. Cohen, and A. Hilliard. 2001. Detective Story: The Airborne Solvent That Contaminated a Water Supply. *OPFLOW*, 27(9): 8-10, 21.

Centers for Disease Control and Prevention. 1999. Critical Biological Agents for Public Health Preparedness. Worksheet.

Clark, R.M., E.E. Geldreich, K.R. Fox, E.W. Rice, C.H. Johnson, J.A. Goodrich, J.A. Barnick, and F. Abdeskan. 1996. Tracking a *Salmonella Serovar Typhi*murium Outbreak in Gideon, Missouri: Role of Contaminant Propagation Modelling. *Aqua Journal of Water Supply Research and Technology*, 45(4):171-183.

Como, J.P. 1983. California Survey of Solvents Leaching From Cold-Applied Coal Tar Paints Used as Internal Coatings in Potable Water Storage Tanks. In *Proc. Of California-Nevada Section AWWA*. Anaheim, CA.

Correia, L. 2002. City of Fall River, Massachusetts. Personal Communication. 508.324.2723.

Environmental Protection Agency. 1999. Guidelines for the Certification and Recertification of the Operators of Community and Nontransient Noncommunity Public Water Systems. *Federal Register*, 64 (24).

Gauthier, V. B. Barbeau, M-C Besner, R. Millette, and M. Prevost. 2000. Storage Tanks Management to Improve Quality of Distributed Water: A Case Study. Extended Abstract for the AWWA International Distribution Systems Research Symposium, Denver, Colo.: AWWA.

Grayman, W.M. and G.J. Kirmeyer. 2000. Quality of Water in Storage. Chapter 11 in *Water Distribution Systems Handbook*, Edited by L.W. Mays, New York. NY: McGraw-Hill.

Grayman, W., L.A. Rossman, C. Arnold, R.A. Deininger, C. Smith, J.F. Smith and R. Schnipke. 2000. *Water Quality Modeling of Distribution System Storage Facilities*. Denver, Colo.: AWWA and AwwaRF.

Great Lakes Upper Mississippi River Board of State Public Health & Environmental Managers. 1997. *Recommended Standards for Water Works. (Also known as Ten States Standards)*. Albany, NY: Health Education Services.

Prepared by AWWA with assistance from Economic and Engineering Services, Inc.

Hickman, D.C. 1999. A Chemical and Biological Warfare Threat: USAF Water Systems at Risk. The Counter Proliferation Papers Future Warfare Series No. 3. USAF Counterproliferation Center, Air War College, Air University, Maxwell Air Force Base, Alabama.

Hunt, T. 2002. Personal Communication with K. Martel. 207-443-2391.

Irias, W.J. 2000. Steel Tank Reservoirs – Replacement vs. Rehabilitation. In *Proc. of the AWWA Infrastructure Conference*. Denver, Colo.: AWWA.

Jacobs, K.A. 2000. Protecting Your Storage Tank – An Analysis of Long- and Short-Term Options. In *Proc. of the AWWA Infrastructure Conference*. Denver, Colo.: AWWA.

Kirmeyer, G.J., L. Kirby, B.M. Murphy, P.F. Noran, K.D. Martel, T.W. Lund, J.L. Anderson, and R. Medhurst. 1999. *Maintaining and Operating Finished Water Storage Facilities*. Denver, Colo.: AWWA and AWWARF.

Kramer, M.H., B.L. Herwaldt, G.F. Craun, R.L. Calderon, and D.D. Juranek. 1996. Waterborne Disease: 1993 and 1994. *JAWWA*, 88(3):66-80.

Krasner, S.W.; Means, E.G. 1985. Returning a Newly Hypalon-Covered Finished Water Reservoir to Service: Health and Aesthetic Considerations. In *Proc. of the AWWA Annual Conference*. Denver, Colo.:AWWA.

Levy, R.V., F.L. Hart, and R.D. Cheetham. 1986. Occurrence and Public Health Significance of Invertebrates in Drinking Water Systems. *JAWWA*, 83(7): 60-66.

Lund, T. 2002. Extech, LLC, Chester, CT. Personal Communication with K. Martel. 860-526-2610.

National Sanitation Foundation. 1996. NSF Standard 61 Drinking Water System Components—Health Effects. ANSI/NSF Standard 61. National Sanitation Foundation International.

Ogg, J.E., R.A. Ryder, and H.L. Smith. 1989. Isolation of *Vibrio Cholerae* From Aquatic Birds in Colorado and Utah. *Applied and Environmental Microbiology*, 55(1):95-99.

Pontius, F.W. 2000. Chloroform: Science, Policy, and Politics. *JAWWA*, 92 (5):12, 14, 16, 18. Reimann-Philipp, U. 2002. Improving Drinking Water Quality in a Small Oklahoma Town: The Dustin Case Study Report. Norman, OK: Floran Technologies. Internal Report.

Smith, C., and G. Burlingame. 1994. Microbial Problems in Treated Water Storage Tanks. In *Proc. of the 1994 Annual AWWA Conference*. Denver, Colo.: AWWA.

U.S. Environmental Protection Agency. 2002a. Chemical Health Effects Tables. http://www.epa.gov/safewater/tcrdsr.html. May 7, 2002.

U.S. Environmental Protection Agency. 2002b. Microbial Health Effects Tables. http://www.epa.gov/safewater/tcrdsr.html. May 7, 2002.

Walski, T.M., 1999. "Modeling TCE dynamics in water distribution tanks". Proc. Water Resource Planning and Management Conf., Reston, Virg.: ASCE.

Walski, T.M. 2000. "Hydraulic Design of Water Distribution Storage Tanks". Chapter 10 in *Water Distribution Systems Handbook*, Edited by L.W. Mays, New York. NY: McGraw-Hill.

Wrc and Public Health Laboratory Service. 1997. Drinking Water Inspectorate Fact Sheets.

Yoo, R.S.; Ellgas, W.M.; Lee, R. 1984. "Water Quality Problems Associated with Reservoir Coatings and Linings." 1984 Annual Conf. Proc. Denver, Colo.:AWWA.

Zelch, C. 2002. Tomcat Consultants, Missouri. Personal Communication with K. Martel. 573-764-5255.

Printed in the United States
By Bookmasters